The Manhattan Project

Making the Atomic Bomb

F. G. Gosling
Office of History and Heritage Resources
Executive Secretariat
Office of Management
Department of Energy

January 2010

National Security History Series

Volume I:
The Manhattan Project: Making the Atomic Bomb

Volume II:
Building the Nuclear Arsenal: Cold War Nuclear Weapons Development
and Production, 1946-1989 (in progress)

Volume III:
Nonproliferation and Stockpile Stewardship:
The Nuclear Weapons Complex in the Post-Cold War World (projected)

The National Security History Series is a joint project of the Office of History and
Heritage Resources and the National Nuclear Security Administration.

Foreword to the 2010 edition

In a national survey at the turn of the millennium, journalists and historians ranked the dropping of the atomic bomb and the surrender of Japan to end the Second World War as the top story of the twentieth century. The advent of nuclear weapons, brought about by the Manhattan Project, not only helped bring an end to World War II but ushered in the atomic age and determined how the next war—the Cold War—would be fought. The Manhattan Project also became the organizational model behind the impressive achievements of American "big science" during the second half of the twentieth century, which demonstrated the relationship between basic scientific research and national security.

This edition of the Office of History's perennial "bestseller" is part of a joint project between the Office of History and Heritage Resources and the National Nuclear Security Administration to produce a three-volume *National Security History Series* documenting the Department of Energy's role in developing, testing, producing, and managing the Nation's nuclear arsenal. There will be hard copy and online versions of each volume as well as complementary websites. The first website–The Manhattan Project: An Interactive History–is available on the Office of History and Heritage Resources website, http://www.cfo. doe.gov/me70/history. The Office of History and Heritage Resources and the National Nuclear Security Administration hope that the *National Security History Series* will provide the Department and the public with reliable and useful information about the national security policies and programs of the Department and its predecessor agencies.

Erica De Vos
Director
Office of the Executive Secretariat

Thomas P. D'Agostino
Administrator
National Nuclear Security Administration

F. G. Gosling
Chief Historian
Office of History and Heritage Resources

Table of Contents

Introduction: The Einstein Letter vii

Part I: Physics Background, 1890-1939 1

Part II: Early Government Support 6

Part III: The Manhattan Engineer District 15

Part IV: The Manhattan Engineer District in Operation 20

Manhattan Project Photo Gallery ....36 - 73

Part V: The Atomic Bomb and American Strategy 87

Part VI: The Manhattan District in Peacetime 99

Manhattan Project Chart 103

Notes 104

Select Bibliography... 106

Manhattan Project Chronology 107

Introduction

Introduction: The Einstein Letter

On October 11, 1939, Alexander Sachs, Wall Street economist and longtime friend and unofficial advisor to President Franklin Delano Roosevelt, met with the President to discuss a letter written by Albert Einstein the previous August. Einstein had written to inform Roosevelt that recent research on chain reactions utilizing uranium made it probable that large amounts of power could be produced by a chain reaction and that, by harnessing this power, the construction of "extremely powerful bombs" was conceivable.[1] Einstein believed the German government was actively supporting research in this area and urged the United States government to do likewise. Sachs read from a cover letter he had prepared and briefed Roosevelt on the main points contained in Einstein's letter. Initially the President was noncommittal and expressed concern over locating the necessary funds, but at a second meeting over breakfast the next morning Roosevelt became convinced of the value of exploring atomic energy.

Einstein drafted his famous letter with the help of the Hungarian émigré physicist Leo Szilard, one of a number of European scientists who had fled to the United States in the 1930s to escape Nazi and Fascist repression. Szilard was among the most vocal of those advocating a program to develop bombs based on recent findings in nuclear physics and chemistry. Those like Szilard and fellow Hungarian refugee physicists Edward Teller and Eugene Wigner regarded it as their responsibility to alert Americans to the possibility that German scientists might win the race to build an atomic bomb and to warn that Hitler would be more than willing to resort to such a weapon. But Roosevelt, preoccupied with events in Europe, took over two months to meet with Sachs after receiving Einstein's letter. Szilard and his colleagues interpreted Roosevelt's inaction as unwelcome evidence that the President did not take the threat of nuclear warfare seriously.

Roosevelt wrote Einstein back on October 19, 1939, informing the physicist that he had set up a committee consisting of civilian and military representatives to study uranium.[2] Events proved that the President was a man of considerable action once he had chosen a direction. In fact, Roosevelt's approval of uranium research in October 1939, based on his belief that the United States could not take the risk of allowing Hitler to achieve unilateral possession of "extremely powerful bombs," was merely the first decision among many that ultimately led to the establishment of the only atomic bomb effort that succeeded in World War II—the Manhattan Project.

The French, who did important research on fission and the feasibility of chain reactions using uranium in 1939 and early 1940, fell under German occupation in June 1940. The British, who made significant theoretical contributions early in the war, did not have the resources to pursue a full-fledged atomic bomb research program while fighting for their survival. Consequently, the British acceded, reluctantly, to American leadership and sent scientists to every Manhattan Project facility. The Germans, despite Allied fears that were not dispelled until the ALSOS mission in 1944,[3] were little nearer to producing atomic weapons at the end of the war than they had been at the beginning of the war. German scientists pursued research on fission, but the government's attempts to forge a coherent strategy met with little success.[4]

The Russians built a program that grew increasingly active as the war drew to a conclusion, but the first successful Soviet test was not conducted until 1949. The Japanese managed to build several cyclotrons by war's end, but the atomic bomb research effort could not maintain a high priority in the face of increasing scarcities. Only the Americans, late entrants into World War II and protected by oceans on both sides, managed to take the discovery of fission from the laboratory to the battlefield and gain a shortlived atomic monopoly.

Part I:
Physics Background, 1890-1939

The Atomic Solar System

The modern effort to uncover the inner structure of the atom began with the discovery of the electron by the English physicist J. J. Thomson in 1897. Thomson proved that cathode rays were not some sort of undefined process occurring in "ether" but were in fact composed of extremely small, negatively charged particles. Dubbed electrons, their exact charge and mass were soon determined by John Townsend and Robert Millikan.

At the same time, discoveries relating to the curious phenomenon of radioactivity had also begun to propel atomic research forward. In 1896, the French physicist Antoine Becquerel detected the three basic forms of radioactivity, which were soon named alpha, beta, and gamma by Ernest Rutherford, a student of Thomson from New Zealand. Also in 1896, the husband-and-wife team of Marie and Pierre Curie began work in Paris on the emission of radiation by uranium and thorium. The Curies soon announced their discoveries of radium and polonium. They also proved that beta particles were negatively charged. In 1900, Becquerel realized that beta particles and electrons were the same things.

In the first decade of the 20th century, Rutherford began to pull all of this information into a coherent whole. In 1903, he proposed that radioactivity was caused by the breakdown of atoms." In 1908, he correctly identified alpha particles as being the nucleus of helium; and in 1911, along with the German physicist Hans Geiger, Rutherford postulated that electrons orbit an atom's nucleus, much as the planets orbit the sun. The second fundamental atomic particle, the proton, was "identified by Rutherford in 1919."

The Danish physicist Neils Bohr combined Rutherford's atomic concepts with Max Planck's quantum theory to produce the first modern model of the atom. In 1913, Bohr demonstrated that electrons moved around an atom's nucleus in certain discrete energy "shells" and that radiation is emitted or absorbed when an electron moves from one shell to another. The following year, Henry Moseley, an English physicist, showed that each element could be identified by its "unique atomic number."

By the 1910s, then, scientists investigating the inner structure of the atom had come to believe, among other things, that energy exists within the atom. Considered in light of Albert Einstein's 1905 theoretical formula $E=mc^2$ (energy equals mass times the square of the velocity of light) stating that matter and energy were equivalent, this belief held breathtaking possibilities. For if Einstein were correct that matter and energy were different forms of the same thing, it followed that anyone unlocking the secrets of how these minute particles were held together–and how they could be broken apart–could produce a massive release of energy. By the late 1910s, then, the stage was set to begin attempting to artificially transmute one atom into another. And if Einstein was correct that matter and energy were different forms of the same thing, it followed that the conversion of a significant quantity of matter into energy should result in a massive release of energy.

The Road to the Bomb

The road to the atomic bomb began in 1919 when Rutherford, working in the Cavendish Laboratory at Cambridge University in England, achieved the first artificial transmutation of an element when he changed several atoms of nitrogen into oxygen. At the time of Rutherford's breakthrough, the atom was conceived as a miniature solar system, with extremely light negatively charged particles, called electrons, in orbit around the much heavier positively charged nucleus. In the process of changing nitrogen into oxygen, Rutherford detected a high-energy particle with a positive charge that proved to be a hydrogen nucleus. The proton, as this subatomic particle was named, joined the electron in the miniature solar system. Another addition came in 1932 when James Chadwick, Rutherford's colleague at Cambridge, identified a third particle, the neutron, so-named because it had no charge.

By the early 1930s, the atom was thought to consist of a positively charged nucleus, containing both protons and neutrons, circled by negatively charged electrons equal in number to the protons in the nucleus. The number of protons deter-mined the element's atomic number. Hydrogen, with one proton, came first and uranium, with ninety-two protons, last on the periodic table. This simple scheme became more complicated when chemists discovered that many elements existed at different weights even while displaying identical chemical properties. It was Chadwick's discovery of the neutron in 1932 that explained this mystery. Scientists found that the weight discrepancy between atoms of the same element resulted because they contained different numbers of neutrons. These different classes of atoms of the same element but with varying numbers of neu-trons were designated isotopes. The three isotopes of uranium, for instance, all have ninety-two protons in their nuclei and ninety-two electrons in orbit. But uranium-238, which accounts for over ninety-nine percent of natural uranium, has 146 neutrons in its nucleus, compared with 143 neutrons in the rare uranium-235 (.7 percent of natural uranium) and 142 neutrons in uranium-234, which is found only in traces in the heavy metal. The slight difference in atomic weight between the uranium-235 and uranium-238

isotopes figured greatly in nuclear physics during the 1930s and 1940s.

The year 1932 produced other notable events in atomic physics. The Englishman J. D. Cockroft and the Irishman E. T. S. Walton, working jointly at the Cavendish Laboratory, were the first to split the atom when they bombarded lithium with protons generated by a particle accelerator and changed the resulting lithium nucleus into two helium nuclei. Also in that year, Ernest O. Lawrence and his colleagues M. Stanley Livingston and Milton White successfully operated the first cyclotron on the Berkeley campus of the University of California.

Moonshine

Lawrence's cyclotron, the Cockroft-Walton machine, and the Van de Graaff electrostatic generator, developed by Robert J. Van de Graaff at Princeton University, were particle accelerators designed to bombard the nuclei of various elements to disintegrate atoms. Attempts of the early 1930s, however, required huge amounts of energy to split atoms because the first accelerators used proton beams and alpha particles as sources of energy. Since protons and alpha particles are positively charged, they met substantial resistance from the positively charged target nucleus when they attempted to penetrate atoms. Even high-speed protons and alpha particles scored direct hits on a nucleus only approximately once in a million tries. Most simply passed by the target nucleus. Not surprisingly, Ernest Rutherford, Albert Einstein, and Niels Bohr regarded particle bombardment as useful in furthering knowledge of nuclear physics but believed it unlikely to meet public expectations of harnessing the power of the atom for practical purposes anytime in the near future. In a 1933 interview Rutherford called such expectations "moonshine."[5] Einstein compared particle bombardment with shooting in the dark at scarce birds, while Bohr agreed that the chances of taming atomic energy were remote.[6]

From Protons to Neutrons: Fermi

Rutherford, Einstein, and Bohr proved to be wrong in this instance, and the proof was not long in coming. Beginning in 1934, the Italian physicist Enrico Fermi began bombarding elements with neutrons instead of protons, theorizing that Chadwick's uncharged particles could pass into the nucleus without resistance. Like other scientists at the time, Fermi paid little attention to the possibility that matter might disappear during bombardment and result in the release of huge amounts of energy in accordance with Einstein's formula, $E=mc^2$, which stated that mass and energy were equivalent. Fermi and his collegues bombarded sixty-three stable elements and produced thirty-seven new radioactive ones.[7] They also found that carbon and hydrogen proved useful as moderators in slowing the bombarding neutrons and that slow neutrons produced the best results since neutrons moving more slowly remained in the vicinity of the nucleus longer and were therefore more likely to be captured.

One element Fermi bombarded with slow neutrons was uranium, the heaviest of the known elements. Scientists disagreed over what Fermi had produced in this transmutation. Some thought that the resulting substances were new "transuranic" elements, while others noted that the chemical properties of the substances resembled those of lighter elements. Fermi was himself uncertain. For the next several years, attempts to identify these substances dominated the research agenda in the international scientific community, with the answer coming out of Nazi Germany just before Christmas 1938.

The Discovery of Fission: Hahn and Strassmann

The radiochemists Otto Hahn and Fritz Strassmann were bombarding elements with neutrons in their Berlin laboratory when they made an unexpected discovery. They found that while the nuclei of most elements changed somewhat during neutron bombardment, uranium nuclei changed greatly and broke into two roughly equal pieces. They split and became not the new transuranic elements that some thought Fermi had discovered but radioactive barium isotopes (barium has the atomic number 56) and fragments of the uranium itself. The substances Fermi had created in his experiments, that is, did more than resemble lighter elements; they were lighter elements. Importantly, the products of the Hahn-Strassmann experiment weighed less than that of the original uranium nucleus, and herein lay the primary significance of their findings. For it followed from Einstein's equation that the loss of mass resulting from the splitting process must have been converted into energy in the form of kinetic energy that could in turn be converted into heat. Calculations made by Hahn's former colleague, Lise Meitner, a refugee from Nazism then staying in Sweden, and her nephew, Otto Frisch, led to the conclusion that so much energy had been released that a previously undiscovered kind of process was at work. Frisch, borrowing the term for cell division in biology—binary fission—named the process fission.[8] For his part, Fermi had produced fission in 1934 but had not recognized it.

Chain Reaction

It soon became clear that the process of fission discovered by Hahn and Strassmann had another important characteristic besides the immediate release of enormous amounts of energy. This was the emission of neutrons. The energy released when fission occurred in uranium caused several neutrons to "boil off" the two main fragments as they flew apart. Given the right set of circumstances, perhaps these secondary neutrons might collide with other atoms and release more neutrons, in turn smashing into other atoms and, at the same time, continuously emitting energy. Beginning with a single uranium nucleus, fission could not only produce substantial amounts of energy but could also lead to a reaction creating

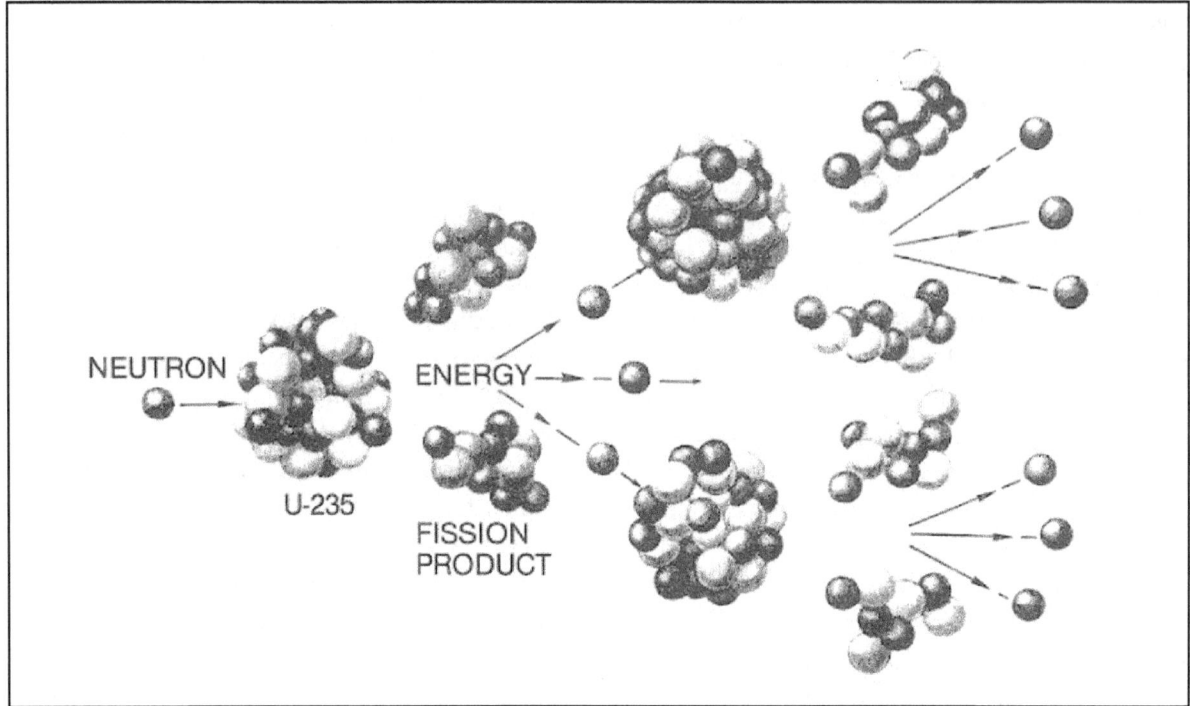

Uranium-235 Fission Chain Reaction. *Department of Energy.*

ever-increasing amounts of energy. The possibility of such a "chain reaction" completely altered the prospects for releasing the energy stored in the nucleus. A controlled self-sustaining reaction could make it possible to generate a large amount of energy for heat and power, while an unchecked reaction could create an explosion of huge force.

Fission Comes to America: 1939

News of the Hahn-Strassmann experiments and the Meitner-Frisch calculations spread rapidly. Meitner and Frisch communicated their results to Niels Bohr, who was in Copenhagen preparing to depart for the United States via Sweden and England. Bohr confirmed the validity of the findings while sailing to New York City, arriving on January 16, 1939. Ten days later Bohr, accompanied by Fermi, communicated the latest developments to some European émigré scientists who had preceded him to this country and to members of the American scientific community at the opening session of a conference on theoretical physics in Washington, D.C.

American physicists quickly grasped the importance of Bohr's message, having by the 1930s developed into an accomplished scientific community. While involved in important theoretical work, Americans made their most significant contributions in experimental physics, where teamwork had replaced individualism in laboratory research. No one epitomized the "can do" attitude of American physicists better than Ernest O. Lawrence, whose ingenuity and drive made the Berkeley Radiation Laboratory the unofficial capital of nuclear physics in the United States. Lawrence staked his claim to American leadership when he built his first particle accelerator, the cyclotron, in 1930. Van de Graaff followed with his generator in 1931, and from then on Americans led the way in producing equipment for nuclear physics and high-energy physics research later.

Early American Work on Fission

American scientists became active participants in attempts to confirm and extend Hahn's and

Strassmann's results, which dominated nuclear physics in 1939. Bohr and John A. Wheeler advanced the theory of fission in important theoretical work done at Princeton University, while Fermi and Szilard collaborated with Walter H. Zinn and Herbert L. Anderson at Columbia University in investigating the possibility of producing a nuclear chain reaction. Given that uranium emitted neutrons (usually two) when it fissioned, the question became whether or not a chain reaction in uranium was possible, and, if so, in which of the three isotopes of the rare metal it was most likely to occur. By March 1940 John R. Dunning and his colleagues at Columbia University, collaborating with Alfred Nier of the University of Minnesota, had demonstrated conclusively that uranium-235, present in only 1 in 140 parts of natural uranium, was the isotope that fissioned with slow neutrons, not the more abundant uranium-238 as Fermi had guessed. This finding was important, for it meant that a chain reaction using the slightly lighter uranium-235 was

possible, but only if the isotope could be separated from the uranium-238 and concentrated into a critical mass, a process that posed serious problems. Fermi continued to try to achieve a chain reaction using large amounts of natural uranium in a pile formation.

Dunning's and Nier's demonstration promised nuclear power but not necessarily a bomb. It was already known that a bomb would require fission by fast neutrons; a chain reaction using slow neutrons might not proceed very far before the metal would blow itself apart, causing little, if any, damage. Uranium-238 fissioned with fast neutrons but could not sustain a chain reaction because it required neutrons with higher energy. The crucial question was whether uranium-235 could fission with fast neutrons in a chain-reacting manner, but without enriched samples of uranium-235, scientists could not perform the necessary experiments.

Part II:
Early Government Support

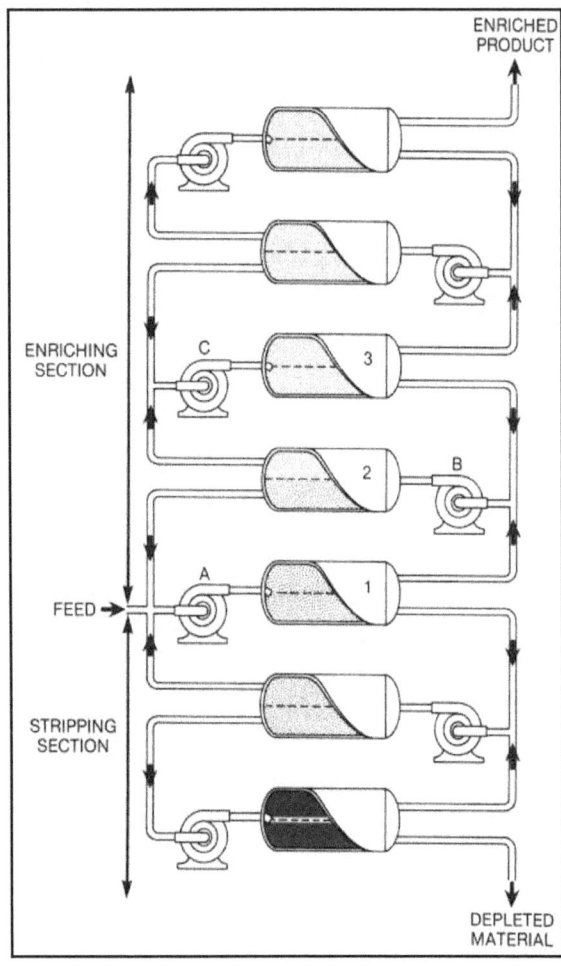

Schematic Diagram of Flow of Process Gas in Gaseous Diffusion Cascade. Reprinted from Richard G. Hewlett and Oscar E. Anderson, Jr., *The New World, 1939-1946*, Volume I of *A History of the United States Atomic Energy Commission* (University Park: Pennsylvania State University Press, 1962).

The Uranium Committee

President Roosevelt responded to the call for government support of uranium research quickly but cautiously. He appointed Lyman J. Briggs, director of the National Bureau of Standards, head of the Advisory Committee on Uranium, which met for the first time on October 21, 1939. The committee, including both civilian and military representation, was to coordinate its activities with Sachs and look into the current state of research on uranium to recommend an appropriate role for the federal government. In early 1940 the Uranium Committee recommended that the government fund limited research on isotope separation as well as Fermi's and Szilard's work on chain reactions at Columbia.

Isotope Separation

Scientists had concluded that enriched samples of uranium-235 were necessary for further research and that the isotope might serve as a fuel source for an explosive device; thus, finding the most effective method of isotope separation was a high priority. Since uranium-235 and uranium-238 were chemically identical, they could not be separated by chemical means. And with their masses differing by less than one percent, separation by physical means would be extremely difficult and expensive. Nonetheless, scientists pressed forward on several complicated techniques of physical separation, all based on the small difference in atomic weight between the uranium isotopes.

The Electromagnetic Method

The electromagnetic method, pioneered by Alfred O.C. Nier of the University of Minnesota, used a mass spectrometer, or spectrograph, to send a stream of charged particles through a magnetic field. Atoms of the lighter isotope would be deflected more by the magnetic field than those of the heavier isotope, resulting in two streams that could then be collected in different receivers. The electromagnetic method as it existed in 1940, however, would have taken far too long to separate quantities sufficient to be useful in the current war. In fact, twenty-seven thousand years would have been required for a single spectrometer to separate one gram of uranium-235.[9]

Gaseous Diffusion

Gaseous diffusion appeared more promising. Based on the well-known principle that molecules of a lighter isotope would pass through a porous barrier more readily than molecules of a heavier one, this approach proposed to produce by myriad repetitions a gas increasingly rich in uranium-235 as the heavier uranium-238 was separated out in a system of cascades. Theoretically, this process could achieve high concentrations of uranium-235 but, like the electromagnetic method, would be extremely costly. British researchers led the way on gaseous diffusion, with John R. Dunning and his colleagues at Columbia University joining the effort in late 1940.

Centrifuge

Many scientists initially thought the best hope for isotope separation was the high-speed centrifuge, a device based on the same principle as the cream separator. Centrifugal force in a cylinder spinning rapidly on its vertical axis would separate a gaseous mixture of two isotopes since the lighter isotope would be less affected by the action and could be drawn off at the center and top of the cylinder. A cascade system composed of hundreds, perhaps thousands, of centrifuges could produce a rich mixture. This method, being pursued primarily by Jesse W. Beams at the University of Virginia, received much of the early isotope separation funding.[10]

Liquid Thermal Diffusion

The Uranium Committee briefly demonstrated an interest in a fourth enrichment process during 1940, only to conclude that it would not be worth pursuing. This process, liquid thermal diffusion, was being investigated by Philip Abelson at the Carnegie Institution. Into the space between two concentric vertical pipes Abelson placed pressurized liquid uranium hexafluoride. With the outer wall cooled by a circulating water jacket and the inner heated by high-pressure steam, the lighter isotope tended to concentrate near the hot wall and the heavier near the cold. Convection would in time carry the lighter isotope to the top of the column. Taller columns would produce more separation. Like other enrichment methods, liquid thermal diffusion was at an early stage of development.[11]

Limited Government Funding: 1940

The Uranium Committee's first report, issued on November 1, 1939, recommended that, despite the uncertainty of success, the government should immediately obtain four tons of graphite and fifty tons of uranium oxide. This recommendation led to the first outlay of government funds—$6,000 in February 1940—and reflected the importance attached to the Fermi-Szilard pile experiments already underway at Columbia University. Building upon the work performed in 1934 demonstrating the value of moderators in producing slow neutrons, Fermi thought that a mixture of the right moderator and natural uranium could produce a self-sustaining chain reaction. Fermi and Szilard increasingly focused their attention on carbon in the form of graphite. Perhaps graphite could slow down, or moderate, the neutrons coming from the fission reaction, increasing the probability of their causing additional fissions in sustaining the chain reaction. A pile containing a large amount of natural uranium could then produce enough secondary neutrons to keep a reaction going.

There was, however, a large theoretical gap between building a self-generating pile and building a bomb. Although the pile envisioned by Fermi and Szilard could produce large amounts of power and might have military applications (powering naval vessels, for instance), it would be too big for a bomb. It would take separation of uranium-235 or substantial enrichment of natural uranium with uranium-235 to create a fast-neutron reaction on a small enough scale to build a usable bomb. While certain of the chances of success in his graphite power pile, Fermi, in 1939, thought that there was "little likelihood of an atomic bomb, little proof that we were not pursuing a chimera."[12]

The National Defense Research Committee

Shortly after World War II began with the German invasion of Poland on September 1, 1939, Vannevar Bush, president of the Carnegie Foundation, became convinced of the need for the government to marshall the forces of science for a war that would inevitably involve the United States. He sounded out other science administrators in the nation's capital and agreed to act as point man in convincing the Roosevelt administration to set up a national science organization. Bush struck an alliance with Roosevelt's closest advisor, Harry Hopkins, and after clearing his project with the armed forces

and science agencies, met with the President and Hopkins. With the imminent fall of France undoubtedly on Roosevelt's mind, it took less than ten minutes for Bush to obtain the President's approval and move into action.[13]

Roosevelt approved in June 1940 the establishment of a voice for the scientific community within the executive branch. The National Defense Research Committee, with Bush at its head, reorganized the Uranium Committee into a scientific body and eliminated military membership. Not dependent on the military for funds, as the Uranium Committee had been, the National Defense Research Committee would have more influence and more direct access to money for nuclear research. In the interest of security, Bush barred foreign-born scientists from committee membership and blocked the further publication of articles on uranium research. Retaining programmatic responsibilities

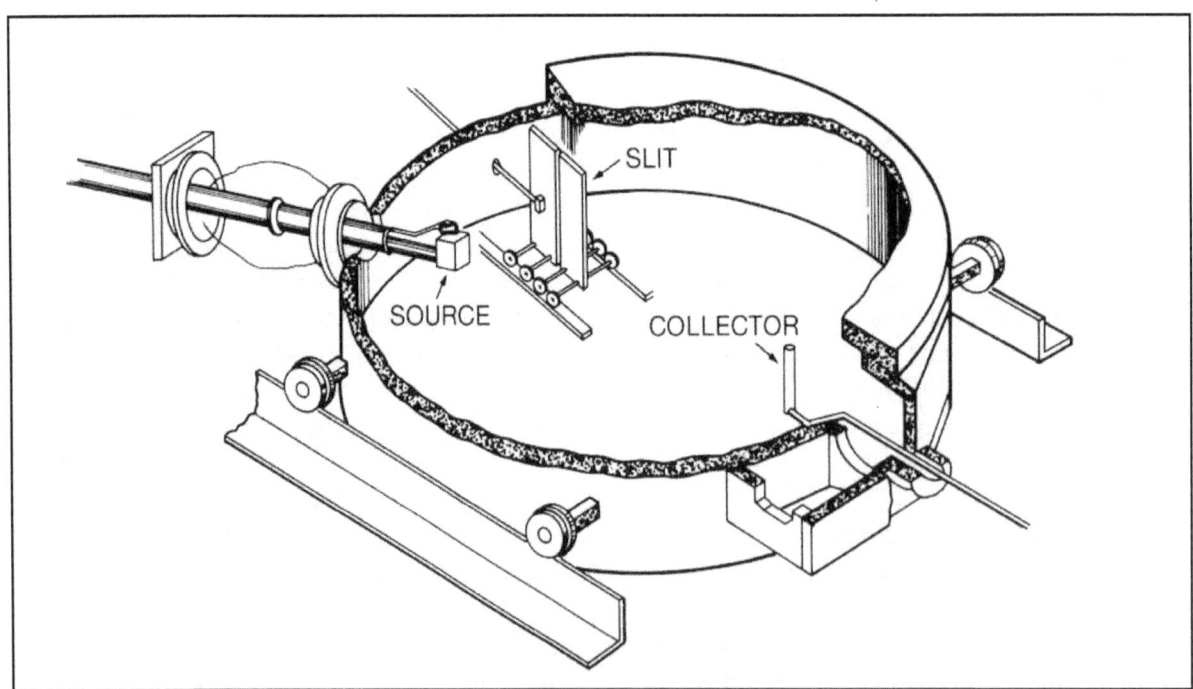

First Mass Spectrograph Components in 37-Inch Cyclotron Tank. Reprinted from Richard G. Hewlett and Oscar E. Anderson, Jr., *The New World, 1939-1946,* Volume I of *A History of the United States Atomic Energy Commission* (University Park: Pennsylvania State University Press, 1962).

for uranium research in the new organizational setup (among the National Defense Research Committee's early priorities were studies on radar, proximity fuzes, and anti-submarine warfare), the Uranium Committee recommended that isotope separation methods and the chain reaction work continue to receive funding for the remainder of 1940. Bush approved the plan and allocated the funds.

A Push From Lawrence

During 1939 and 1940, most of the work done on isotope separation and the chain reaction pile was performed in university laboratories by academic scientists funded primarily by private foundations. While the federal government began supporting uranium research in 1940, the pace appeared too leisurely to the scientific community and failed to convince scientists that their work

was of high priority. Certainly few were more inclined to this view than Ernest O. Lawrence, director of the Radiation Laboratory at the University of California in Berkeley. Lawrence was among those who thought that it was merely a matter of time before the United States was drawn into World War II, and he wanted the government to mobilize its scientific forces as rapidly as possible.

Specifically what Lawrence had on his mind in early 1941 were experiments taking place in his own laboratory using samples produced in the cyclotron. Studies on uranium fission fragments by Edwin M. McMillan and Philip H. Abelson led to the chemical identification of element 93, neptunium, while research by Glenn T. Seaborg revealed that an isotope of neptunium decayed to yet another transuranium (man-made) element. In

Ernest Lawrence, Arthur Compton, Vannevar Bush, James Conant, Karl Compton. and Alfred Loomis. Reprinted from Richard G. Hewlett and Oscar E. Anderson, Jr., *The New World, 1939-1946,* Volume I of *A History of the United States Atomic Energy Commission* (University Park: Pennsylvania State University Press, 1962).

February, Seaborg identified this as element 94, which he later named plutonium. By May he had proven that plutonium-239 was 1.7 times as likely as uranium-235 to fission. This finding made the Fermi-Szilard experiment more important than ever as it suggested the possibility of producing large amounts of the fissionable plutonium in a uranium pile using plentiful uranium-238 and then separating it chemically. Surely this would be less expensive and simpler than building isotope-separation plants.

Lawrence, demonstrating his characteristic energy and impatience, launched a campaign to speed up uranium research. He began by proposing to convert his smaller cyclotron into a spectrograph to produce uranium-235. Since both the cyclotron and the spectrograph used a vacuum chamber and electromagnet, this conversion would be relatively uncomplicated. Lawrence then took his case to Karl T. Compton and Alfred L. Loomis at Harvard University, both doing radar work for the National Defense Research Committee and benefiting from Lawrence's advice in staffing their laboratories. Infected by Lawrence's enthusiasm, Compton forwarded Lawrence's optimistic assessment on uranium research to Bush, warning that Germany was undoubtedly making progress and that Briggs and the Uranium Committee were moving too slowly. Compton also noted that the British were ahead of their American colleagues, even though, in his opinion, they were inferior in both numbers and ability.

Program Review: Summer 1941

Bush and Lawrence met in New York City. Though he continued to support the Uranium Committee, Bush recognized that Lawrence's assessment was not far off the mark. Bush shrewdly decided to appoint Lawrence as an advisor to Briggs—a move that quickly resulted in funding for plutonium work at Berkeley and for Nier's mass spectrograph at Minnesota—and also asked the National Academy of Sciences

to review the uranium research program. Headed by Arthur Compton of the University of Chicago and including Lawrence, this committee submitted its unanimous report on May 17. Compton's committee, however, failed to provide the practical-minded Bush with the evidence he needed that uranium research would pay off in the event the United States went to war in the near future. Compton's group thought that increased uranium funding could produce radioactive material that could be dropped on an enemy by 1943, a pile that could power naval vessels in three or four years, and a bomb of enormous power at an indeterminate point, but certainly not before 1945. Compton's report discussed bomb production only in connection with slow neutrons, a clear indication that much more scientific work remained to be done before an explosive device could be detonated.[14]

Bush reconstituted the National Academy of Sciences committee and instructed it to assess the recommendations contained in the first report from an engineering standpoint. On July 11, the second committee endorsed the first report and supported continuation of isotope separation work and pile research for scientific reasons, though it admitted that it could promise no immediate applications. The second report, like the first, was a disappointing document from Bush's point of view.[15]

The Office of Scientific Research and Development

By the time Bush received the second National Academy of Sciences report, he had assumed the position of director of the Office of Scientific Research and Development. Established by an executive order on June 28, 1941—six days after German troops invaded the Soviet Union—the Office of Scientific Research and Development strengthened the scientific presence in the federal government. Bush, who had lobbied hard for the new setup, now reported directly to the President and could invoke the prestige of the White House in his dealings with other federal agencies. The National Defense Research Committee, now headed by James B. Conant, president of Harvard University, became an advisory body responsible for making research and development recommendations to the Office of Scientific Research and Development. The Uranium Committee became the Office of Scientific Research and Development Section on Uranium and was codenamed S-l (Section One of the Office of Scientific Research and Development).

Turning the Corner: The MAUD Report

Bush's disappointment with the July 11 National Academy of Sciences report did not last long. Several days later he and Conant received a copy of a draft report forwarded from the National Defense Research Committee liaison office in London. The report, prepared by a group codenamed the MAUD Committee and set up by the British in spring 1940 to study the possibility of developing a nuclear weapon, maintained that a sufficiently purified critical mass of uranium-235 could fission even with fast neutrons.[16] Building upon theoretical work on atomic bombs performed by refugee physicists Rudolf Peierls and Otto Frisch in 1940 and 1941, the MAUD report estimated that a critical mass of ten kilograms would be large enough to produce an enormous explosion. A bomb this size could be loaded on existing aircraft and be ready in approximately two years.[17]

Americans had been in touch with the MAUD Committee since fall 1940, but it was the July 1941 MAUD report that helped the American bomb effort turn the corner. Here were specific plans for producing a bomb, produced by a distinguished group of scientists with high credibility in the United States, not only with Bush and Conant but with the President.[18] The MAUD report dismissed plutonium production, thermal diffusion, the electromagnetic method, and the centrifuge and called for gaseous diffusion of uranium-235 on a massive scale. The British believed that uranium research could lead to the production of a bomb in time to effect the outcome of the war. While the MAUD report provided encouragement to Americans advocating a more extensive uranium research program, it also served as a sobering reminder that fission had been discovered in Nazi Germany almost three years earlier and that since spring 1940 a large part of the Kaiser Wilhelm Institute in Berlin had been set aside for uranium research.

Bush and Conant immediately went to work. After strengthening the Uranium Committee, particularly with the addition of Fermi as head of theoretical studies and Harold C. Urey as head of isotope separation and heavy water research (heavy water was highly regarded as a moderator), Bush asked yet another reconstituted National Academy of Sciences committee to evaluate the uranium program. This time he gave Compton specific instructions to address technical questions of critical mass and destructive capability, partially to verify the MAUD results.

Bush Reports to Roosevelt

Without waiting for Compton's committee to finish its work, Bush went to see the President. On October 9 Bush met with Roosevelt and Vice President Henry A. Wallace (briefed on uranium research in July). Bush summarized the British findings, discussed cost and duration of a bomb project, and emphasized the uncertainty of the situation. He also received the President's permission to explore construction needs with the Army. Roosevelt instructed him to move as quickly as possible but not to go beyond research and development. Bush, then, was to find out if a bomb could be built and at what cost but not to proceed to the production stage without further presidential authorization. Roosevelt indicated that he could find a way to finance the project and asked Bush to draft a letter so that the British government could be approached "at the top."[19]

Compton reported back on November 6, just one month and a day before the Japanese attack on Pearl Harbor on December 7, 1941, brought the United States into World War II (Germany and Italy declared war on the United States three days later). Compton's committee concluded that a critical mass of between two and 100 kilograms of uranium-235 would produce a powerful fission bomb and that for $50-100 million isotope separation in sufficient quantities could be accomplished. Although the Americans were less optimistic than the British, they confirmed the basic conclusions of the MAUD committee and convinced Bush to forward their findings to Roosevelt under a cover letter on November 27. Roosevelt did not respond until January 19, 1942; when he did, it was as commander in chief of a nation at war. The President's handwritten note read, "V. B. OK—returned—I think you had best keep this in your own safe FDR."[20]

Moving Into Action

By the time Roosevelt responded, Bush had set the wheels in motion. He put Eger V. Murphree, a chemical engineer with the Standard Oil Company, in charge of a group responsible for overseeing engineering studies and supervising pilot plant construction and any laboratory-scale investigations. And he appointed Urey, Lawrence, and Compton as program chiefs. Urey headed up work including diffusion and centrifuge methods and heavy-water studies. Lawrence took electromagnetic and plutonium responsibilities, and Compton ran chain reaction and weapon theory programs. Bush's responsibility was to coordinate engineering and scientific efforts and make final decisions on recommendations for construction contracts. In accordance with the instructions he received from Roosevelt, Bush removed all uranium work from the National Defense Research Committee. From this point forward, broad policy decisions relating to uranium were primarily the responsibility of the Top Policy Group, composed of Bush, Conant,

Vice President Wallace, Secretary of War Henry L. Stimson, and Army Chief of Staff George C. Marshall.[21] A high-level conference convened by Wallace on December 16 put the seal of approval on these arrangements. Two days later the S-1 Committee gave Lawrence $400,000 to continue his electromagnetic work.

With the United States now at war and with the fear that the American bomb effort was behind Nazi Germany's, a sense of urgency permeated the federal government's science enterprise. Even as Bush tried to fine-tune the organizational apparatus, new scientific information poured in from laboratories to be analyzed and incorporated into planning for the upcoming design and construction stage. By spring 1942, as American naval forces slowed the Japanese advance in the Pacific with an April victory in the battle of the Coral Sea, the situation had changed from one of too little money and no deadlines to one of a clear goal, plenty of money, but too little time. The race for the bomb was on.

Continuing Efforts on Isotope Separation

During the first half of 1942, several routes to a bomb were explored. At Columbia, Urey worked on the gaseous diffusion and centrifuge systems for isotope separation in the codenamed SAM (Substitute or Special Alloy Metals) Laboratory. At Berkeley, Lawrence continued his investigations on electromagnetic separation using the mass spectrograph he had converted from his thirty-seven-inch cyclotron. Compton patched together facilities at the University of Chicago's Metallurgical Laboratory for pile experiments aimed at producing plutonium. Meanwhile Murphree's group hurriedly studied ways to move from laboratory experiments to production facilities.

Research on uranium required uranium ore, and obtaining sufficient supplies was the responsibility of Murphree and his group. Fortunately, enough ore was on hand to meet the projected need of 150 tons through mid-1944. Twelve hundred tons of high-grade ore were stored on Staten Island, and Murphree made arrangements to obtain additional supplies from Canada and the Colorado Plateau, the only American source. Uranium in the form of hexafluoride was also needed as feed material for the centrifuge and the gaseous and thermal diffusion processes. Abelson, who had moved from the Carnegie Institution to the Naval Research Laboratory, was producing small quantities, and Murphree made arrangements with E. I. du Pont de Nemours and Company and the Harshaw Chemical Company of Cleveland to produce hexafluoride on a scale sufficient to keep the vital isotope separation research going.

Lawrence was so successful in producing enriched samples of uranium-235 electromagnetically with his converted cyclotron that Bush sent a special progress report to Roosevelt on March 9, 1942. Bush told the President that Lawrence's work might lead to a short cut to the bomb, especially in light of new calculations indicating that the critical mass required might well be smaller than previously predicted. Bush also emphasized that the efficiency of the weapon would probably be greater than earlier estimated and expressed more confidence that it could be detonated successfully. Bush thought that if matters were expedited a bomb was possible in 1944. Two days later the President responded: "I think the whole thing should be pushed not only in regard to development, but also with due regard to time. This is very much of the essence."[22]

In the meantime, however, isotope separation studies at Columbia quickly confronted serious engineering difficulties. Not only were the specifications for the centrifuge demanding, but, depending upon rotor size, it was estimated that it would require tens of thousands of centrifuges

to produce enough uranium-235 to be of value. Gaseous diffusion immediately ran into trouble as well. Fabrication of an effective barrier to separate the uranium isotopes seemed so difficult as to relegate gaseous diffusion to a lower priority (the barrier had to be a corrosion-resistant membrane containing millions of submicroscopic holes per square inch). Both separation methods demanded the design and construction of new technologies and required that parts, many of them never before produced, be finished to tolerances not previously imposed on American industry.

In Chicago, Compton decided to combine all pile research by stages. Initially he funded Fermi's pile at Columbia and the theoretical work of Eugene Wigner at Princeton and J. Robert Oppenheimer at Berkeley. He appointed Szilard head of materials acquisition and arranged for Seaborg to move his plutonium work from Berkeley to Chicago in April 1942. Compton secured space wherever he could find it, including a racket court under the west grandstand at Stagg Field, where Samuel K. Allison began building a graphite and uranium pile. Although it was recognized that heavy water would provide a moderator superior to graphite, the only available supply was a small amount that the British had smuggled out of France. In a decision typical of the new climate of urgency, Compton decided to forge ahead with graphite, a decision made easier by Fermi's increasingly satisfactory results at Columbia and Allison's even better results in Chicago. In light of recent calculations that cast doubt on the MAUD report's negative assessment of plutonium production, Compton hoped that Allison's pile would provide plutonium that could be used as material for a weapon.

By May 1942, Bush decided that production planning could wait no longer, and he instructed Conant to meet with the S-1 section leaders and make recommendations on all approaches to the bomb, regardless of cost. Analyzing the status of the four processes then under consideration for producing fissionable materials for a bomb—the gaseous diffusion, centrifuge, and electromagnetic uranium isotope separation methods and the plutonium producing pile—the committee decided on May 23 to recommend that all four be pushed as fast as possible to the pilot plant stage and to full production planning. This decision reflected the inability of the committee to distinguish a clear front-runner and its consequent unwillingness to abandon any method. With funds readily available and the outcome of the war conceivably hanging in the balance, the S-l leadership recommended that all four methods proceed to the pilot plant stage and to full production planning.

Enter the Army

The decision to proceed with production planning led directly to the involvement of the Army, specifically the Corps of Engineers. Roosevelt had approved Army involvement on October 9, 1941, and Bush had arranged for Army participation at S-l meetings beginning in March 1942. The need for security suggested placing the S-l program within one of the armed forces, and the construction expertise of the Corps of Engineers made it the logical choice to build the production facilities

envisioned in the Conant report of May 23. By orchestrating some delicate negotiations between the Office of Scientific Research and Development and the Army, Bush was able to transfer the responsibility for process development, materials procurement, engineering design, and site selection to the Corps of Engineers and to earmark approximately sixty percent of the proposed 1943 budget, or $54 million, for these functions. An Army officer would be in overall command of the entire project. This new arrangement left S-l, with a budget of approximately $30 million, in charge of only university research and pilot plant studies. Additional reorganization created an S-l Executive Committee, composed of Conant, Briggs, Compton, Lawrence, Murphree, and Urey. This group would oversee all Office of Scientific Research and Development work and keep abreast of technical developments that might influence engineering considerations or plant design.[23] With this reorganization in place, the nature of the American atomic bomb effort changed from one dominated by research scientists to one in which scientists played a supporting role in the construction enterprise run by the United States Army Corps of Engineers.

Part III:
The Manhattan Engineer District

Initial Problems

Summer 1942—during which the American island-hopping campaign in the Pacific began at Guadalcanal—proved to be a troublesome one for the fledgling bomb project. Colonel James C. Marshall received the assignment of directing the Laboratory for the Development of Substitute Metals, or DSM. Marshall immediately moved from Syracuse to New York City, where he set up the Manhattan Engineer District, established by general order on August 13. Marshall, like most other Army officers, knew nothing of nuclear physics. Furthermore, Marshall and his Army superiors were disposed to move cautiously. In one case, for instance, Marshall delayed purchase of an excellent production site in Tennessee pending further study, while the scientists who had been involved in the project from the start were pressing for immediate purchase. While Bush had carefully managed the transition to Army control, there was not yet a mechanism to arbitrate disagreements between S-1 and the military. The resulting lack of coordination complicated attempts to gain a higher priority for scarce materials and boded ill for the future of the entire bomb project.

Reorganization of the Manhattan Engineer District:
Groves and the Military Policy Committee

Decisions made in September provided administrative clarity and renewed the project's sense of urgency. Bush and the Army agreed that an officer other than Marshall should be given the assignment of overseeing the entire atomic project, which by now was referred to as the Manhattan Project. On September 17, the Army appointed Colonel Leslie R. Groves (promoted to Brigadier General six days later) to head the effort. Groves was an engineer with impressive credentials, including building of the Pentagon, and, most importantly, had strong administrative abilities. Within two days Groves acted to obtain the Tennessee site and secured a higher priority rating for project materials. In addition, Groves moved the Manhattan Engineer District headquarters from New York to Washington. He quickly recognized the talents of Marshall's deputy, Colonel Kenneth D. Nichols, and arranged for Nichols to work as his chief aide and troubleshooter throughout the war.

Bush, with the help and authority of Secretary of War Henry L. Stimson, set up the Military Policy Committee, including one representative each from the Army, the Navy, and the Office of Scientific Research and Development. Bush hoped

General Leslie R. Groves. Reprinted from Vincent C. Jones, *Manhattan: The Army and the Atomic Bomb* (Washington, D.C.: U.S. Government Printing Office, 1985).

that scientists would have better access to decision making in the new structure than they had enjoyed when DSM and S-1 operated as parallel but separate units. With Groves in overall command (Marshall remained as District Engineer, where his cautious nature proved useful in later decision making) and the Military Policy Committee in place (the Top Policy Group retained broad policy authority), Bush felt that early organizational deficiencies had been remedied.[24]

During summer and fall 1942, technical and administrative difficulties were still severe. Each of the four processes for producing fissionable material for a bomb remained under consideration, but a full-scale commitment to all four posed serious problems even with the project's high priority. When Groves took

J. Robert Oppenheimer. Reprinted by Permission of the J. Robert Oppenheimer Memorial Committee.

command in mid-September, he made it clear that by late 1942 decisions would be made as to which process or processes promised to produce a bomb in the shortest amount of time. The exigencies of war, Groves held, required scientists to move from laboratory research to development and production in record time. Though traditional scientific caution might be short-circuited in the process, there was no alternative if a bomb was to be built in time to be used in the current conflict. As everyone involved in the Manhattan Project soon learned, Groves never lost sight of this goal and made all his decisions accordingly.

Producing Fissionable Materials: Fall 1942

Groves made good on his timetable when he scheduled a meeting of the Military Policy Committee on November 12 and a meeting of the S-1 Executive Committee on November 14. The scientists at each of the institutions doing isotope separation research knew these meetings would determine the separation method to be used in the bomb project; therefore, the keen competition among the institutions added to the sense of urgency created by the war. Berkeley remained a hotbed of activity as Lawrence and his staff pushed the electromagnetic method into the lead. The S-1 Executive Committee even toyed with the idea of placing all its money on Lawrence but was dissuaded by Conant. Throughout the summer and fall, Lawrence refined his new 184-inch magnet and huge cyclotron to produce calutrons, as the tanks were called in honor of the University of California, capable of reliable beam resolution and containing improved collectors for trapping the enriched uranium-235. The S-1 Executive Committee visited Berkeley on September 13 and subsequently recommended building both a pilot plant and a large section of a full-scale plant in Tennessee.

The centrifuge being developed by Jesse Beams at the University of Virginia was the big loser in the November meetings. Westinghouse had been unable to overcome problems with its model centrifuge. Parts failed with discouraging regularity due to severe vibrations during trial runs; consequently, a pilot plant and subsequent production stages appeared impractical in the near future. Conant had already concluded that the centrifuge was likely to be dropped when he reported to Bush on October 26. The meetings of November 12 and 14 confirmed his analysis.

Gaseous diffusion held some promise and remained a live option, although the Dunning group at Columbia had not yet produced any uranium-235 by the November meetings. The major problem continued to be the barrier; nickel was the leading candidate for barrier material, but there was serious doubt as to whether a reliable nickel barrier could be ready in sufficient quantity by the end of the war.

While the centrifuge was cancelled and gaseous diffusion received mixed reviews, optimism prevailed among the pile proponents at the Metallurgical Laboratory in Chicago. Shortages of uranium and graphite delayed construction of the Stagg Field pile—CP-1 (Chicago Pile Number One)—but this frustration was tempered by calculations indicating that a completed pile would produce a chain reaction. With Fermi's move to Chicago in April, all pile research was now being conducted at the Metallurgical Laboratory as Compton had planned, and Fermi and his team anticipated a successful experiment by the end of the year. Further optimism stemmed from Seaborg's inventive work with plutonium, particularly his investigations on plutonium's oxidation states that seemed to provide a way to separate plutonium from the irradiated uranium to be produced in the pile. In August Seaborg's team produced a microscopic sample of pure plutonium, a major chemical achievement and one fully justifying further work on the pile. The only cloud in the Chicago sky was the scientists' disappointment when they learned that

construction and operation of the production facilities, now to be built near the Clinch River in Tennessee at Site X, would be turned over to a private firm. An experimental pile would be built in the Argonne Forest Preserve just outside Chicago, but the Metallurgical Laboratory scientists would have to cede their claim to pile technology to an organization experienced enough to take the process into construction and operation.

The Luminaries Report From Berkeley

While each of the four processes fought to demonstrate its "workability" during summer and fall 1942, equally important theoretical studies were being conducted that greatly influenced the decisions made in November. Robert Oppenheimer headed the work of a group of theoretical physicists he called the luminaries, which included Felix Bloch, Hans Bethe, Edward Teller, and Robert Serber, while John H. Manley assisted him by coordinating nationwide fission research and instrument and measurement studies from the Metallurgical Laboratory in Chicago. Despite inconsistent experimental results, the consensus emerging at Berkeley was that approximately twice as much fissionable material would be required for a bomb than had been estimated six months earlier. This was disturbing, especially in light of the military's view that it would take more than one bomb to win the war. The goal of mass-producing fissionable material, which still appeared questionable in late 1942, seemed even more unrealistic given Oppenheimer's estimates. Oppenheimer did report, with some enthusiasm, that fusion explosions using deuterium (heavy hydrogen) might be possible. The possibility of thermonuclear (fusion) bombs generated some optimism since deuterium supplies, while not abundant, were certainly larger and more easily supplemented than were those of uranium and plutonium. S-1 immediately authorized basic research on other light elements.

Input From DuPont

Final input for the November meetings of the Military Policy Committee and the S-1 Executive Committee came from DuPont. One of the first things Groves did when he took over in September was to begin courting DuPont, hoping that the giant chemical firm would undertake construction and operation of the plutonium separation plant to be built in Tennessee. He appealed to patriotism, informing the company that the bomb project had high priority with the President and maintaining that a successful effort could affect the outcome of the war. DuPont managers resisted but did not refuse the task, and in the process they provided an objective appraisal of the pile project. Noting that it was not even known if the chain reaction would work, DuPont stated that under the best of circumstances plutonium could be mass-produced by 1945, and it emphasized that it thought the chances of this happening were low. This appraisal did not discourage Groves, who was confident that DuPont would take the assignment if offered.

Time for Decisions

The Military Policy Committee met on November 12, 1942, and its decisions were ratified by the S-1 Executive Committee two days later. The Military Policy Committee, acting on Groves's and Conant's recommendations, cancelled the centrifuge project. Gaseous diffusion, the pile, and the electromagnetic method were to proceed directly to full-scale, eliminating the pilot plant stage. The S-1 Executive Committee approved these recommendations and agreed that the gaseous diffusion facility was of lower priority than either the pile or the electromagnetic plant but ahead of a second pile. The scientific committee also asked DuPont to look into methods for increasing American supplies of heavy water in case it was needed to serve as a moderator for one of the new piles.

A Brief Scare

Anxious as he was to get moving, Groves decided to make one final quality control check before acting on the decisions of November 12 and 14. This decision seemed imperative after a brief scare surrounding the pile project. While Fermi's calculations provided reasonable assurance against such a possibility, the vision of a chain reaction running wild in heavily-populated Chicago arose when the S-1 Executive Committee found that Compton was building the experimental pile at Stagg Field, a decision he had made without informing either the committee or Groves. In addition, information from British scientists raised serious questions about the feasibility of deriving plutonium from the pile. It took several days for Groves and a committee of scientists including Lawrence and Oppenheimer to satisfy themselves that the pile experiment posed little danger, was justified by sound theory, and would in all probability produce plutonium as predicted.

One Last Look: The Lewis Committee

On November 18, Groves appointed Warren K. Lewis of the Massachusetts Institute of Technology to head a final review committee, comprised of himself and three DuPont representatives. During the next two weeks, the committee traveled from New York to Chicago to Berkeley and back again through Chicago. It endorsed the work on gaseous diffusion at Columbia, though it made some organizational recommendations; in fact, the Lewis committee elevated gaseous diffusion to first priority and expressed reservations about the electromagnetic program despite an impassioned presentation by Lawrence in Berkeley. Upon returning to Chicago, Crawford H. Greenewalt, a member of the Lewis committee, was present at Stagg Field when, at 3:20 p.m. on December 2, 1942, Fermi's massive lattice pile of 400 tons of graphite, six tons of uranium metal, and fifty tons of uranium oxide achieved the first self-sustaining chain reaction, operating initially at a power level of one-half watt (increased to 200 watts ten days later).[25] As Compton reported to Conant, "the Italian navigator has just landed in the new world." To Conant's question, "Were the natives friendly?" Compton answered, "Everyone landed safe and happy."[26] Significant as this moment was in the history of physics, it came after the Lewis committee had endorsed moving to the pilot stage and one day after Groves had instructed DuPont to move into design and construction on December 1.[27]

No Turning Back: Final Decisions and Presidential Approval

The S-l Executive Committee met to consider the Lewis report on December 9, 1942, just weeks after Allied troops landed in North Africa. Most of the morning session was spent evaluating the controversial recommendation that only a small electromagnetic plant be built. Lewis and his colleagues based their recommendation on the belief that Lawrence could not produce enough uranium-235 to be of military significance. But since the calutron could provide enriched samples quickly, the committee supported the construction of a small electromagnetic plant. Conant disagreed with the Lewis committee's assessment, believing that uranium had more weapon potential than plutonium. And since he knew that gaseous diffusion could not provide any enriched uranium until the gaseous diffusion plant was in full operation, he supported the one method that might, if all went well, produce enough uranium to build a bomb in 1944. During the afternoon, the S-l Executive Committee went over a draft Groves had prepared for Bush to send to the President. It supported the Lewis committee's report except that it recommended skipping the pilot plant stage for the pile. After Conant and the Lewis committee met on December 10 and reached a compromise on the electromagnetic method, Groves's draft was amended and forwarded to Bush.[28]

On December 28, 1942, President Roosevelt approved the establishment of what ultimately became a government investment in excess of $2 billion, $.5 billion of which was itemized in Bush's report submitted on December 16. The Manhattan Project was authorized to build full-scale gaseous diffusion and plutonium plants and the compromise electromagnetic plant, as well as heavy water production facilities. In his report, Bush reaffirmed his belief that bombs possibly could be produced during the first half of 1945 but cautioned that an earlier delivery was unlikely. No schedule could guarantee that the United States would overtake Germany in the race for the bomb, but by the beginning of 1943 the Manhattan Project had the complete support of President Roosevelt and the military leadership, the services of some of the nation's most distinguished scientists, and a sense of urgency driven by fear. Much had been achieved in the year between Pearl Harbor and the end of 1942.

No single decision created the American atomic bomb project. Roosevelt's December 28 decision was inevitable in light of numerous earlier ones that, in incremental fashion, committed the United States to pursuing atomic weapons. In fact, the essential pieces were in place when Roosevelt approved Bush's November 9, 1941, report on January 19, 1942. At that time, there was a science organization at the highest level of the federal government and a Top Policy Group with direct access to the President. Funds were authorized, and the participation of the Corps of Engineers had been approved in principle. In addition, the country was at war and its scientific leadership—as well as its President—had the belief, born of the MAUD report, that the project could result in a significant contribution to the war effort. Roosevelt's approval of $500 million in late December 1942 was a step that followed directly from the commitments made in January of that year and stemmed logically from the President's earliest tentative decisions in late 1939.

Part IV:
The Manhattan Engineer District in Operation

The Manhattan Project

In many ways the Manhattan Engineer District operated like any other large construction company. It purchased and prepared sites, let contracts, hired personnel and subcontractors, built and maintained housing and service facilities, placed orders for materials, developed administrative and accounting procedures, and established communications networks. By the end of the war Groves and his staff had spent approximately $2.2 billion on production facilities and towns built in the states of Tennessee, Washington, and New Mexico, as well as on research in university laboratories from Columbia to Berkeley. What made the Manhattan Project unlike other companies performing similar functions was that, because of the necessity of moving quickly, it invested hundreds of millions of dollars in unproven and hitherto unknown processes and did so entirely in secret. Speed and secrecy were the watchwords of the Manhattan Project.

Secrecy proved to be a blessing in disguise. Although it dictated remote site locations, required subterfuge in obtaining labor and supplies, and served as a constant irritant to the academic scientists on the project, it had one overwhelming advantage: Secrecy made it possible to make decisions with little regard for normal peacetime political considerations. Groves knew that as long as he had the backing of the White House money would be available and he could devote his considerable energies entirely to running the bomb project. Secrecy in the Manhattan Project was so complete that many

people working for the organization did not know what they were working on until they heard about the bombing of Hiroshima on the radio. The need for haste clarified priorities and shaped decision making. Unfinished research on three separate, unproven processes had to be used to freeze design plans for production facilities, even though it was recognized that later findings inevitably would dictate changes. The pilot plant stage was eliminated entirely, violating all manufacturing practices and leading to intermittent shutdowns and endless troubleshooting during trial runs in production facilities. The inherent problems of collapsing the stages between the laboratory and full production created an emotionally charged atmosphere with optimism and despair alternating with confusing frequency.

Despite Bush's assertion that a bomb could probably be produced by 1945, he and the other principals associated with the project recognized the magnitude of the task before them. For any large organization to take laboratory research into design, construction, operation, and product delivery in two-and-a-half years (from early 1943 to Hiroshima) would be a major industrial achievement. Whether the Manhattan Project would be able to produce bombs in time to affect the current conflict was an open question as 1943 began. (Obvious though it seems in retrospect, it must be remembered that no one at the time knew that the war would end in 1945 or who the remaining contestants would be if and when the atomic bomb was ready for use.)

Clinton Engineer Works (Oak Ridge)

By the time President Roosevelt authorized the Manhattan Project on December 28, 1942, work on the east Tennessee site where the first production facilities were to be built was already underway. The final quarter of 1942 saw the acquisition of the roughly ninety-square-mile parcel (59,000 acres) in the ridges just west of Knoxville, the removal of the relatively few families on the marginal farmland, and extensive site preparation to provide the transportation, communications, and utility needs of the town and production plants that would occupy the previously underdeveloped area. Original plans called for the Clinton Engineer Works, as the military reservation was named, to house approximately 13,000 people in prefabricated housing, trailers, and wood dormitories. By the time the Manhattan Engineer District headquarters were moved from Washington to Tennessee in the summer of 1943 (Groves kept the Manhattan Project's office in Washington and placed Nichols in command in Tennessee), estimates for the town of Oak Ridge had been revised upward to 40-45,000 people. (The name Oak Ridge did not come into widespread usage until after World War II but will be used here to avoid confusion.) At the end of the war, Oak Ridge was the fifth largest town in Tennessee, and the Clinton Engineer Works was consuming one-seventh of all the power being produced in the nation.[29] While the Army and its contractors tried to keep up with the rapid influx of workers and their families, services always lagged behind demand, though morale remained high in the atomic boomtown.

The three production facility sites were located in valleys away from the town. This provided security and containment in case of explosions. The Y-12 area, home of the electromagnetic plant, was closest to Oak Ridge, being but one ridge away to the south. Farther to the south and

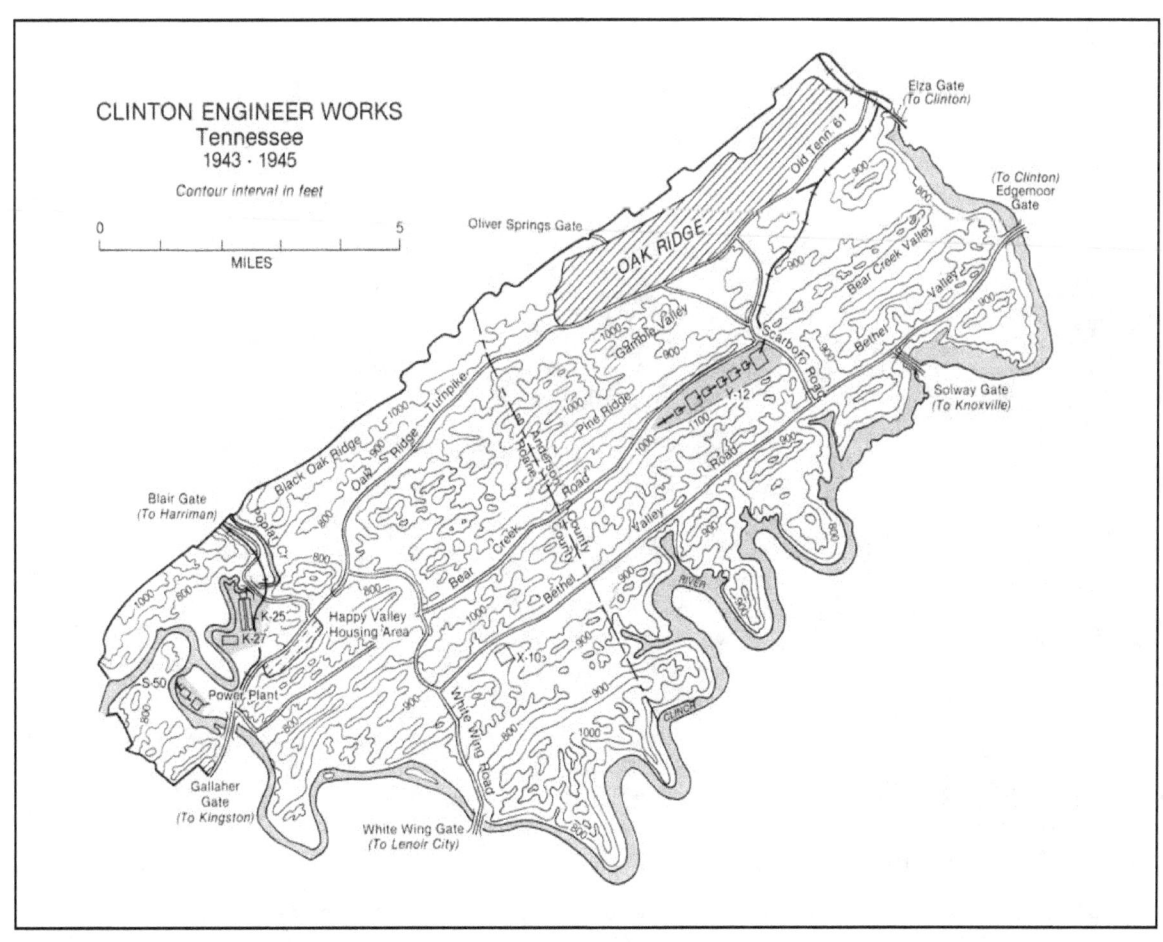

Clinton Engineer Works. Reprinted from Vincent C. Jones, *Manhattan: The Army and the Atomic Bomb* (Washington, D.C.: U.S. Government Printing Office, 1985).

west lay both the X-10 area, which contained the experimental plutonium pile and separation facilities, and K-25, site of the gaseous diffusion plant and later the S-50 thermal diffusion plant. Y-12 and X-10 were begun slightly earlier in 1943 than was K-25, but all three were well along by the end of the year.

The Y-12 Electromagnetic Plant: Final Decisions

Although the Lewis report had placed gaseous diffusion ahead of the electromagnetic approach, many were still betting in early 1943 that Lawrence and his mass spectrograph would

eventually predominate. Lawrence and his laboratory of mechanics at Berkeley continued to experiment with the giant 184-inch magnet, trying to reach a consensus on which shims, sources, and collectors to incorporate into Y-12 design for the Oak Ridge plant. Research on magnet size and placement and beam resolution eventually led to a racetrack configuration of two magnets with forty-eight gaps containing two vacuum tanks each per building, with ten buildings being necessary to provide the 2,000 sources and collectors needed to separate 100 grams of uranium-235 daily. It was hoped that improvements in calutron design, or placing

Y-12 Alpha Racetrack at Clinton. Spare vacuum tanks in Left Foreground. Reprinted from Richard G. Hewlett and Oscar E. Anderson, Jr., *The New World, 1939-1946,* Volume I of *A History of the United States Atomic Energy Commission* (University Park: Pennsylvania State University Press, 1962).

multiple sources and collectors in each tank, might increase efficiency and reduce the number of tanks and buildings required, but experimental results were inconclusive even as Stone & Webster of Boston, the Y-12 contractor at Oak Ridge, prepared to break ground.

At a meeting of Groves, Lawrence, and John R. Lotz of Stone & Webster in Berkeley late in December 1942, Y-12 plans took shape. It was agreed that Stone & Webster would take over design and construction of a 500-tank facility, while Lawrence's laboratory would play a supporting role by supplying experimental data.

By the time another summit conference on Y-12 took place in Berkeley on January 13 and 14, Groves had persuaded the Tennessee Eastman Corporation to sign on as plant operator and arranged for various parts of the electromagnetic equipment to be manufactured by the Westinghouse Electric and Manufacturing Company, the Allis-Chalmers Manufacturing Company, and the Chapman Valve Manufacturing Company. General Electric agreed to provide electrical equipment.

On January 14, after a day of presentations and a demonstration of the experimental tanks in

Y-12 Electromagnetic Plant Under Construction at Clinton. Reprinted from Richard G. Hewlett and Oscar E. Anderson, Jr., *The New World, 1939-1946*, Volume I of *A History of the United States Atomic Energy Commission* (University Park: Pennsylvania State University Press, 1962).

the cyclotron building, Groves stunned the Y-12 contractors by insisting that the first racetrack of ninety-six tanks be in operation by July 1 and that 500 tanks be delivered by year's end. Given that each racetrack was 122 feet long, 77 feet wide and 15 feet high; that the completed plant was to be the size of three two-story buildings; that tank design was still in flux; and that chemical extraction facilities also would have to be built, Groves's demands were little less than shocking. Nonetheless, Groves maintained that his schedule could be met.[30]

For the next two months Lawrence, the contractors, and the Army negotiated over the final design. While all involved could see possible improvements, there simply was not enough time to incorporate every suggested modification. Y-12 design was finalized at a March 17 meeting in Boston, with one major modification—the inclusion of a second stage of the electromagnetic process. The purpose of this second stage was to take the enriched uranium-235 derived from several runs of the first stage and use it as the sole feed material for a second stage of racetracks containing tanks approximately half

Y-12 Beta Racetrack at Clinton. Reprinted from Richard G. Hewlett and Oscar E. Anderson, Jr., *The New World, 1939-1946*, Volume I of *A History of the United States Atomic Energy Commission* (University Park: Pennsylvania State University Press, 1962).

the size of those in the first. Groves approved this arrangement and work began on both the Alpha (first-stage) and Beta (second-stage) tracks.

Construction of Y-12

Groundbreaking for the Alpha plant took place on February 18, 1943. Soon blueprints could not be produced fast enough to keep up with construction as Stone & Webster labored to meet Groves's deadline. The Beta facility was actually begun before formal authorization. While laborers were aggressively recruited, there was always a shortage of workers skilled enough to perform jobs according to the rigid specifications. (A further complication was that some tasks could be performed only by workers with special clearances.) Huge amounts of material had to be obtained (38 million board feet of lumber, for instance), and the magnets needed so much copper for windings that the Army had to borrow almost 15,000 tons of silver bullion from the United States Treasury to fabricate into strips and wind on to coils as a substitute for copper.[31] Treasury silver was also used to manufacture the busbars that ran around the top of the racetracks.

Replacing copper with silver solved the immediate problem of the magnets and busbars, but persistent shortages of electronic tubes, generators, regulators, and other equipment plagued the electromagnetic project and posed the most serious threat to Groves's deadline. Furthermore, last-minute design changes continued to frustrate equipment manufacturers. Nonetheless, when Lawrence toured with Y-12 contractors in May 1943, he was impressed by the scale of operations. Lawrence returned to Berkeley rededicated to the "awful job" of finishing the racetracks on time.[32]

Design Changes at Y-12

Lawrence and his colleagues continued to look for ways to improve the electromagnetic process. Lawrence found that hot (high positive voltage) electrical sources could replace the single cold (grounded) source in future plants, providing more efficient use of power, reducing insulator failure, and making it possible to use multiple rather than single beams.[33] Meanwhile, receiver design evolved quickly enough in spring and summer 1943 to be incorporated into the Alpha plant. Work at the Radiation Laboratory picked up additional speed in March with the authorization of the Beta process. With Alpha technology far from perfected, Lawrence and his staff now had to participate in planning for an unanticipated stage of the electromagnetic process.

While the scientists in Berkeley studied changes that would be required in the down-sized Beta racetracks, engineering work at Oak Ridge prescribed specific design modifications. For a variety of reasons, including simplicity of maintenance, Tennessee Eastman decided that the Beta plant would consist of two tracks of thirty-six tanks each in a rectangular, rather than oval, arrangement. Factoring this configuration into their calculations, Lawrence and his coworkers bent their efforts to developing chemical processing techniques that would minimize the loss of enriched uranium during Beta production runs. To make certain that Alpha had enough feed material, Lawrence arranged for research on an alternate method at Brown University and expanded efforts at Berkeley. With what was left of his time and money in early 1943 Lawrence built prototypes of Alpha and Beta units for testing and training operating personnel. Meanwhile Tennessee Eastman, running behind schedule, raced to complete experimental models so that training and test runs could be performed at Oak Ridge.

Warning From Los Alamos

But in the midst of encouraging progress in construction and research on the electromagnetic process in July came discouraging news from Oppenheimer's isolated laboratory in Los Alamos, New Mexico, set up in spring 1943 to consolidate work on atomic weapons. Oppenheimer warned that three times more fissionable material would

be required for a bomb than earlier estimates had indicated. Even with satisfactory performance of the racetracks, it was possible that they might not produce enough purified uranium-235 in time. Lawrence responded to this crisis in characteristic fashion: He immediately lobbied Groves to incorporate multiple sources into the racetracks under construction and to build more racetracks. Groves decided to build the first four as planned but, after receiving favorable reports from both Stone & Webster and Tennessee Eastman, allowed a four-beam source in the fifth. Convinced that the electromagnetic process would work and sensing that estimates from Los Alamos might be revised downward in the future, Groves let Lawrence talk him into building a new plant—the Y-12 Extension—doubling the size of the electromagnetic complex. The Alpha component of the Y-12 Extension was designated as Alpha II and would consist of two buildings, each with two rectangular racetracks of ninety-six tanks operating with four-beam sources. Also authorized was a second Beta building containing two racetracks. Improvisation remained the key work at Oak Ridge.

Shakedown at Y-12

During summer and fall 1943, the first electromagnetic plant began to take shape. The huge building to house the operating equipment was readied as manufacturers began delivering everything from electrical switches to motors, valves, and tanks. While construction and outfitting proceeded, almost 5,000 operating and maintenance personnel were hired and trained. Then, between October and mid-December, Y-12 paid the price for being a new technology that had not been put through its paces in a pilot plant. Vacuum tanks in the first Alpha racetrack leaked and shimmied out of line due to magnetic pressure, welds failed, electrical circuits malfunctioned, and operators made frequent mistakes. Most seriously, the magnet coils shorted out because of rust and sediment in the cooling oil.

Groves arrived on December 15 and shut the racetrack down. The coils were sent to Allis-Chalmers with hope that they could be cleaned without being dismantled entirely, while measures were taken to prevent recurrence of the shorting problem. The second Alpha track now bore the weight of the electromagnetic effort. In spite of precautions aimed at correcting the electrical and oil-related problems that had shut down the first racetrack, the second Alpha fared little better when it started up in mid-January 1944. While all tanks operated at least for short periods, performance was sporadic and maintenance could not keep up with electrical failures and defective parts. Like its predecessor, Alpha 2 was a maintenance nightmare.

Alpha 2 produced about 200 grams of twelve-percent uranium-235 by the end of February, enough to send samples to Los Alamos and feed the first Beta unit but not enough to satisfy estimates of weapon requirements. The first four Alpha tracks did not operate together until April, a full four months late. While maintenance improved, output was well under previous expectations. The opening of the Beta building on March 11 led to further disappointment. Beam resolution was so unsatisfactory that complete redesign was required. To make matters worse, word spread that the K-25 gaseous diffusion process was in deep trouble because of its ongoing barrier crisis. K-25 had been counted upon to provide uranium enriched enough to serve as feed material for Beta. Now it would be producing such slight enrichment that the Alpha tracks would have to process K-25's material, requiring extensive redesign and retooling of tanks, doors, and liners, particularly in units that would be wired to run as hot, rather than as cold, electrical sources.[34]

Reworking the Racetracks

It became clear to Groves that he would have to find a way for a combination of isotope separation processes to produce enough fissionable material for bombs. This meant making changes in the racetracks so that they could process the slightly enriched material produced by K-25. He then concentrated on further expansion of the electromagnetic facilities. Lawrence, seconded by Oppenheimer, believed that four more racetracks should be built to accompany the nine already finished or under construction. Groves agreed with this approach, though he was not sure that the additional racetracks could be built in time.

As K-25 stock continued to drop and plutonium prospects remained uncertain, Lawrence lobbied yet again for further expansion of Y-12, arguing that it provided the only possible avenue to a bomb by 1945. His plan was to convert all tanks to multiple beams and to build two more racetracks. By this time even the British had given up on gaseous diffusion and urged acceptance of Lawrence's plan.

Time was running out, and an element of desperation crept into decisions made at a meeting on July 4, 1944. Groves met with the Oak Ridge contractors to consider proposals Lawrence had prepared after assessing once again the resources and abilities of the Radiation Laboratory. There was to be no change in the completed racetracks; there simply was not enough time. Some improvements were to be made in the racetracks then under construction. In the most important decision made at the meeting, Lawrence was to throw all he had into a completely new type of calutron that would use a thirty-beam source. Technical support would come from both Westinghouse and General Electric, which would cease work on four-beam development. It was a gamble in a high-stakes game, but sticking with the Alpha and Beta racetracks might have been an even greater gamble.

K-25 Gaseous Diffusion Plant Under Construction at Clinton. Reprinted from Richard G. Hewlett and Oscar E. Anderson, Jr., *The New World, 1939-1946*, Volume I of *A History of the United States Atomic Energy Commission* (University Park: Pennsylvania State University Press, 1962).

K-25 from Opposite End. White Building in Center of Previous Picture Discernible at Far End. Reprinted from Richard G. Hewlett and Oscar E. Anderson, Jr., *The New World, 1939-1946,* Volume I of *A History of the United States Atomic Energy Commission* (University Park: Pennsylvania State University Press, 1962).

The K-25 Gaseous Diffusion Plant

Eleven miles southwest of Oak Ridge on the Clinch River was the site of the K-25 gaseous diffusion plant upon which so much hope had rested when it was authorized in late 1942. Championed by the British and placed first by the Lewis committee, gaseous diffusion seemed to be based on sound theory but had not yet produced samples of enriched uranium-235.

At Oak Ridge, on a relatively flat area of about 5,000 acres, site preparation for the K-25 power-plant began in June. Throughout the summer, contractors contended with primitive roads as they shipped in the materials needed to build what became the world's largest steam electric plant. In September work began on the cascade building, plans for which had changed dramati-

cally since the spring. Now there were to be fifty four-story buildings (2,000,000 square feet) in a U-shape measuring half a mile by 1,000 feet. Innovative foundation techniques were required to avoid setting thousands of concrete piers to support load-bearing walls.

Since it was eleven miles from the headquarters at Oak Ridge, the K-25 site developed into a satellite town. Housing was supplied, as was a full array of service facilities for the population that reached 15,000. Dubbed Happy Valley by the inhabitants, the town had housing similar to that in Oak Ridge, but, like headquarters, it too experienced chronic shortages. Even with a contractor camp with facilities for 2,000 employees nearby, half of Happy Valley's workers had to commute to the construction site daily.

Downgrading K-25

In late summer 1943, it was decided that K-25 would play a lesser role than originally intended. Instead of producing fully enriched uranium-235, the gaseous diffusion plant would now provide around fifty percent enrichment for use as feed material in Y-12. This would be accomplished by eliminating the more troublesome upper part of the cascade. Even this level of enrichment was not assured since a barrier for the diffusion plant still did not exist. The decision to downgrade K-25 was part of the larger decision to double Y-12 capacity and fit with Groves's new strategy of utilizing a combination of methods to produce enough fissionable material for bombs as soon as possible.

There was no doubt in Groves's mind that gaseous diffusion still had to be pursued vigorously. Not only had major resources already been expended on the program, but there was also the possibility that it might yet prove successful. Y-12 was in trouble as 1944 began, and the plutonium pile projects were just getting underway. A workable barrier design might put K-25 ahead in the race for the bomb. Unfortunately, no one had been able to fabricate barrier of sufficient quality. The only alternative remaining was to increase production enough to compensate for the low percentage of barrier that met specifications. As Lawrence prepared to throw everything he had into a thirty-beam source for Y-12, Groves ordered a crash barrier program, hoping to prevent K-25 from standing idle as the race for the bomb continued.

Help From the Navy

As problems with both Y-12 and K-25 reached crisis proportions in spring and summer 1944, the Manhattan Project received help from an unexpected source—the United States Navy. President Roosevelt had instructed that the atomic bomb effort be an Army program and that the Navy be excluded from deliberations. Navy research on atomic power, conducted primarily for submarines, received no direct aid from Groves, who, in fact, was not up-to-date on the state of Navy efforts when he received a letter on the subject from Oppenheimer late in April 1944.

Oppenheimer informed Groves that Philip Abelson's experiments on thermal diffusion at the Philadelphia Naval Yard deserved a closer look. Abelson was building a plant to produce enriched uranium to be completed in early July. It might be possible, Oppenheimer thought, to help Abelson complete and expand his plant and use its slightly enriched product as feed for Y-12 until problems with K-25 could be resolved.

The liquid thermal diffusion process had been evaluated in 1940 by the Uranium Committee, when Abelson was at the National Bureau of Standards. In 1941 he moved to the Naval Research Laboratory, where there was more support for his work. During summer 1942 Bush and Conant received reports about Abelson's research but concluded that it would take too long for the thermal diffusion process to make a major contribution to the bomb effort, especially since the electromagnetic and pile projects were making satisfactory progress. After a visit with Abelson in January 1943, Bush encouraged the Navy to increase its support of thermal diffusion. A thorough review of Abelson's project early in 1943, however, concluded that thermal diffusion work should be expanded but should not be considered as a replacement for gaseous diffusion, which was better understood theoretically. Abelson continued his work independently of the Manhattan Project. He obtained authorization to build a new plant at the Philadelphia Naval Yard, where construction began in January 1944.

Groves immediately saw the value of Oppenheimer's suggestion and sent a group to Philadelphia to visit Abelson's plant. A quick analysis demonstrated that a thermal diffusion plant could be built at Oak Ridge and placed in operation by early 1945. The steam needed in the convection columns was already at hand in the form of the almost completed K-25 powerplant. It would be a relatively simple matter to provide steam to the thermal diffusion plant and produce enriched uranium, while providing electricity for the K-25 plant when it was finished. Groves gave the contractor, H. K. Ferguson Company of Cleveland, just ninety days from September 27 to bring a 2,142-column plant on line (Abelson's plant contained 100 columns). There was no time to waste as Happy Valley braced itself for a new influx of workers.

The Metallurgical Laboratory

One of the most important branches of the far-flung Manhattan Project was the Metallurgical Laboratory (Met Lab) in Chicago, which was counted on to design a production pile for plutonium. Here again the job was to design equipment for a technology that was not well understood even in the laboratory. The Fermi pile, important as it was historically, provided little technical guidance other than to suggest a lattice arrangement of graphite and uranium. Any pile producing more power than the few watts generated in Fermi's famous experiment would require elaborate controls, radiation shielding, and a cooling system. These engineering features would all contribute to a reduction in neutron multiplication (neutron multiplication being represented by k); so it was imperative to determine which pile design would be safe and controllable and still have a k high enough to sustain an ongoing reaction.[35]

Pile Design

A group headed by Compton's chief engineer, Thomas V. Moore, began designing the production pile in June 1942. Moore's first goals were to find the best methods of extracting plutonium from the irradiated uranium and for cooling the uranium. It quickly became clear that a production pile would differ significantly in design from Fermi's experimental reactor, possibly by extending uranium rods into and through the graphite next to cooling tubes and building a radiation and containment shield. Although experimental reactors like Fermi's did not generate enough power to need cooling systems, piles built to produce plutonium would operate at high power levels and require coolants. The Met Lab group considered the full range of gases and liquids in a search to isolate the substances with the best nuclear characteristics, with hydrogen and helium standing out among the gases and water—even with its marginal nuclear properties and tendency to corrode uranium—as the best liquid.

During the summer, Moore and his group began planning a helium-cooled pilot pile for the Argonne Forest Preserve near Chicago, built by Stone & Webster, and on September 25 they reported to Compton. The proposal was for a 460-ton cube of graphite to be pierced by 376 vertical columns containing twenty-two cartridges of uranium and graphite. Cooling would be provided by circulating helium from top to bottom through the pile. A wall of graphite surrounding the reactor would provide radiation containment, while a series of spherical segments that gave the design the nickname Mae West would make up the outer shell.

By the time Compton received Moore's report, he had two other pile designs to consider. One was a water-cooled model developed by Eugene Wigner and Gale Young, a former colleague of Compton's. Wigner and Young proposed a twelve-foot by twenty-five-foot cylinder of graphite with pipes of uranium extending from a water tank above, through the cylinder, and into a second water tank underneath. Coolant would circulate continuously through the system, and corrosion would be minimized by coating interior surfaces or lining the uranium pipes.

A second alternative to Mae West was more daring. Szilard thought that liquid metal would be such an efficient coolant that, in combination with an electromagnetic pump having no moving parts (adapted from a design he and Einstein had created), it would be possible to achieve high power levels in a considerably smaller pile. Szilard had trouble obtaining supplies for his experiment, primarily because bismuth, the metal he preferred as the coolant, was rare.

Groves Steps In

October 1942 found Groves in Chicago ready to force a showdown on pile design. Szilard was noisily complaining that decisions had to be made so that design could move to procurement and construction. Compton's delay reflected uncertainty of the superiority of the helium pile and awareness that engineering studies could not be definitive until the precise value of k had been established. Some scientists at the Met Lab urged that a full production pile be built immediately, while others advocated a multi-step process, perhaps beginning with an externally cooled reactor proposed by Fermi. The situation was tailor-made for a man with Groves's temperament. On October 5 Groves exhorted the Met Lab to decide on pile design within a week. Even wrong decisions were better than no decisions, Groves claimed, and since time was more valuable than money, more than one approach should be pursued if no single design stood out. While Groves did not mandate a specific decision, his imposed deadline forced the Met Lab scientists to reach a consensus.

Compton decided on compromise. Fermi would study the fundamentals of pile operation on a small experimental unit to be completed and in

operation by the end of the year. Hopefully he could determine the precise value of k and make a significant advance in pile engineering possible. An intermediate pile with external cooling would be built at Argonne and operated until June 1, 1943, when it would be taken down for plutonium extraction. The helium-cooled Mae West, designed to produce 100 grams of plutonium a day, would be built and operating by March 1944. Studies on liquid-cooled reactors would continue, including Szilard's work on liquid metals.

Seaborg and Plutonium Chemistry

While the Met Lab labored to make headway on pile design, Seaborg and his coworkers tried to gain enough information about transuranium chemistry to insure that plutonium produced could be successfully extracted from the irradiated uranium. Using lanthanum fluoride as a carrier, Seaborg isolated a weighable sample of plutonium in August 1942. At the same time, Isadore Perlman and William J. Knox explored the peroxide method of separation; John E. Willard studied various materials to determine which best adsorbed plutonium;[36] Theodore T. Magel and Daniel K. Koshland, Jr., researched solvent-extraction processes; and Harrison S. Brown and Orville F. Hill performed experiments into volatility reactions. Basic research on plutonium's chemistry continued as did work on radiation and fission products.

Seaborg's discovery and subsequent isolation of plutonium were major events in the history of chemistry, but, like Fermi's achievement, it remained to be seen whether they could be translated into a production process useful to the bomb effort. In fact, Seaborg's challenge seemed even more daunting, for while piles had to be scaled up ten to twenty times, a separation plant for plutonium would involve a scale-up of the laboratory experiment on the order of a billion-fold.

Collaboration with DuPont's Charles M. Cooper and his staff on plutonium separation facilities began even before Seaborg succeeded in isolating a sample of plutonium. Seaborg was reluctant to drop any of the approaches then under consideration, and Cooper agreed. The two decided to pursue all four methods of plutonium separation but put first priority on the lanthanum fluoride process Seaborg had already developed. Cooper's staff ran into problems with the lanthanum fluoride method in late 1942, but by then Seaborg had become interested in phosphate carriers. Work led by Stanley G. Thompson found that bismuth phosphate retained over ninety-eight percent plutonium in a precipitate. With bismuth phosphate as a backup for the lanthanum fluoride, Cooper moved ahead on a semiworks near Stagg Field.

DuPont Joins the Team

Compton's original plans to build the experimental pile and chemical separation plant on the University of Chicago campus changed during fall 1942. The S-1 Executive Committee concurred that it would be safer to put Fermi's pile in Argonne and build the pilot plant and separation facilities in Oak Ridge than to place these experiments in a populous area. On October 3 DuPont agreed to design and build the chemical separation plant. Groves tried to entice further DuPont participation at Oak Ridge by having the firm prepare an appraisal of the pile project and by placing three DuPont staff members on the Lewis committee. Because DuPont was sensitive about its public image (the company was still smarting from charges that it profiteered during World War I), Groves ultimately obtained the services of the giant chemical company for the sum of one dollar over actual costs. In addition, DuPont vowed to stay out of the bomb business after the war and offered all patents to the United States government.

Groves had done well in convincing DuPont to join the Manhattan Project. DuPont's proven administrative structure assured excellent coordination (Crawford Greenewalt was given the responsibility of coordinating DuPont and Met Lab planning), and Groves and Compton welcomed the company's demand that it be put in full charge of the Oak Ridge plutonium project. DuPont had a strong organization and had studied every aspect of the Met Lab's program thoroughly before accepting the assignment. While deeply involved in the overall war effort, DuPont expected to be able to divert personnel and other resources from explosives work in time to throw its full weight into the Oak Ridge project.

Moving the pilot plutonium plant to Oak Ridge left too little room for the full-scale production plant at the X-10 site and also left too little generating power for yet another major facility. Furthermore, the site was uncomfortably close to Knoxville should a catastrophe occur. Thus the search for an alternate location for the full-scale plutonium facility began soon after DuPont joined the production team. Compton's scientists needed an area of approximately 225 square miles. Three or four piles and one or two chemical separation complexes would be at least a mile apart for security purposes, while nothing would be allowed within four miles of the separation complexes for fear of radioactive accidents. Towns, highways, rail lines, and laboratories would be several miles further away.

Hanford

December 16, 1942, found Colonel Franklin T. Matthias of Groves's staff and two DuPont engineers headed for the Pacific Northwest and southern California to investigate possible production sites. Of the possible sites available, none had a better combination of isolation, long construction season, and abundant water for hydroelectric power than those found along the Columbia and Colorado Rivers. After viewing six locations in Washington, Oregon, and California, the group agreed that the area around Hanford, Washington, best met the criteria established by

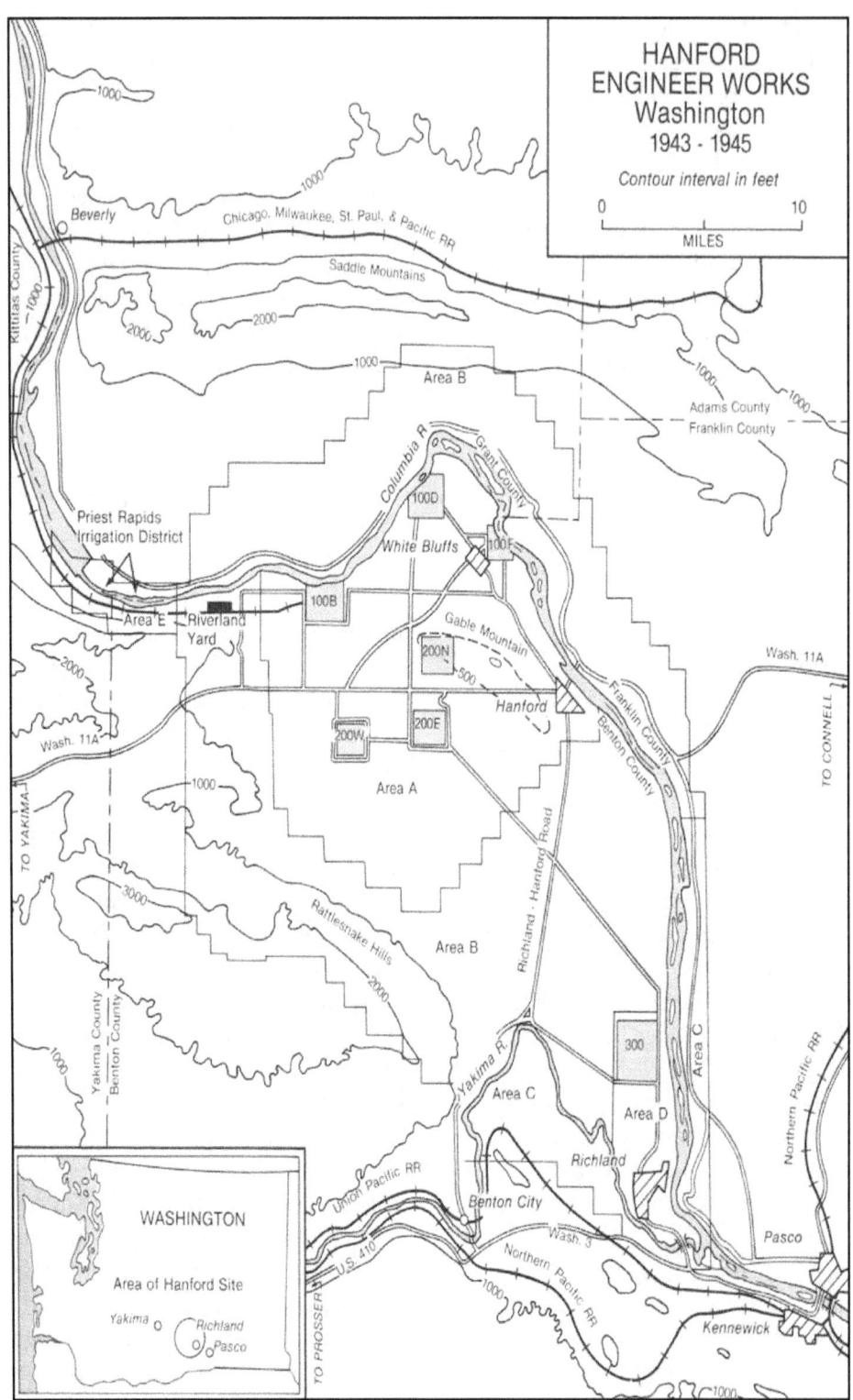

Hanford Engineer Works. Reprinted from Vincent C. Jones, *Manhattan: The Army and the Atomic Bomb* (Washington, D.C.: U.S. Government Printing Office, 1985).

the Met Lab scientists and DuPont engineers. The Grand Coulee and Bonneville Dams offered substantial hydroelectric power, while the flat but rocky terrain would provide excellent support for the huge plutonium production buildings. The ample site of nearly one-half million acres was far enough inland to meet security requirements, while existing transportation facilities could quickly be improved and labor was readily available. Pleased with the committee's unanimous report, Groves accepted its recommendation and authorized the establishment of the Hanford Engineer Works, codenamed Site W.

Now that DuPont would be building the plutonium production complex in the Northwest, Compton saw no reason for any pile facilities in Oak Ridge and proposed to conduct Met Lab research in either Chicago or Argonne. DuPont, on the other hand, continued to support a semiworks at Oak Ridge and asked the Met Lab scientists to operate it. Compton demurred on the grounds that he did not have sufficient technical staff, but he was also reluctant because his scientists complained that their laboratory was becoming little more than a subsidiary of DuPont. In the end, Compton knew the Met Lab would have to support DuPont, which simply did not have sufficient expertise to operate the semiworks on its own. The University of Chicago administration supported Compton's decision in early March.

Pile Design: Changing Priorities

The fall 1942 planning sessions at the Met Lab led to the decision to build a second Fermi pile at Argonne as soon as his experiments on the first were completed and to proceed on design of the Mae West helium-cooled unit. When DuPont engineers assessed the Met Lab's plans in the late fall, they agreed that helium should be given first priority. They placed heavy water second and urged an all-out effort to produce more of this highly effective moderator. Bismuth and water

were ranked third and fourth in DuPont's analysis. Priorities changed when Fermi's calculations demonstrated a higher value for k than anyone had anticipated. Met Lab scientists concluded that a water-cooled pile was now feasible, while DuPont remained interested in helium." Since a helium-cooled unit shared important design characteristics with an air-cooled one, Greenewalt thought that an air-cooled semiworks at Oak Ridge would contribute significantly to designing the full-scale facilities at Hanford.

DuPont established the general specifications for the air-cooled semiworks and chemical separation facilities in early 1943. A massive graphite block, protected by several feet of concrete, would contain hundreds of horizontal channels filled with uranium slugs surrounded by cooling air. New slugs would be pushed into the channels on the face of the pile, forcing irradiated ones at the rear to fall into an underwater bucket. The buckets of irradiated slugs would undergo radioactive decay for several weeks, then be moved by underground canal into the chemical separation facility where the plutonium would be extracted with remote control equipment.

Met Lab activities focused on designing a water-cooled pile for the full-scale plutonium plant. Taking their cue from the DuPont engineers, who utilized a horizontal design for the air-cooled semiworks, Met Lab scientists abandoned the vertical arrangement with water tanks, which had posed serious engineering difficulties. Instead, they proposed to place uranium slugs sealed in aluminum cans inside aluminum tubes. The tubes, laid horizontally through a graphite block, would cool the pile with water injected into each tube. The pile, containing 200 tons of uranium and 1,200 tons of graphite, would need 75,000 gallons of water per minute for cooling.

Decision on Pile Design

Greenewalt's initial response to the water-cooled design was guarded. He worried about pressure problems that might lead to boiling water in individual tubes, corrosion of slugs and tubes, and the one-percent margin of safety for k. But he was even more worried about the proposed helium-cooled model. He feared that the compressors would not be ready in time for Hanford, that the shell could not be made vacuum-tight, and that the pile would be extremely difficult to operate. DuPont engineers conceded that Greenewalt's fears were well-grounded. Late in February, Greenewalt reluctantly concluded that the Met Lab's model, while it had its problems, was superior to DuPont's own helium-cooled design and decided to adopt the water-cooled approach.

The Met Lab's victory in the pile design competition came as its status within the Manhattan Project was changing. Still an exciting place intellectually, the Met Lab occupied a less central place in the bomb project as Oak Ridge and Hanford rose to prominence. Fermi continued to work on the Stagg Field pile (CP-1), hoping to determine the exact value of k. Subsequent experiments at the Argonne site using CP-2, built with material from CP-1, focused on neutron capture probabilities, control systems,

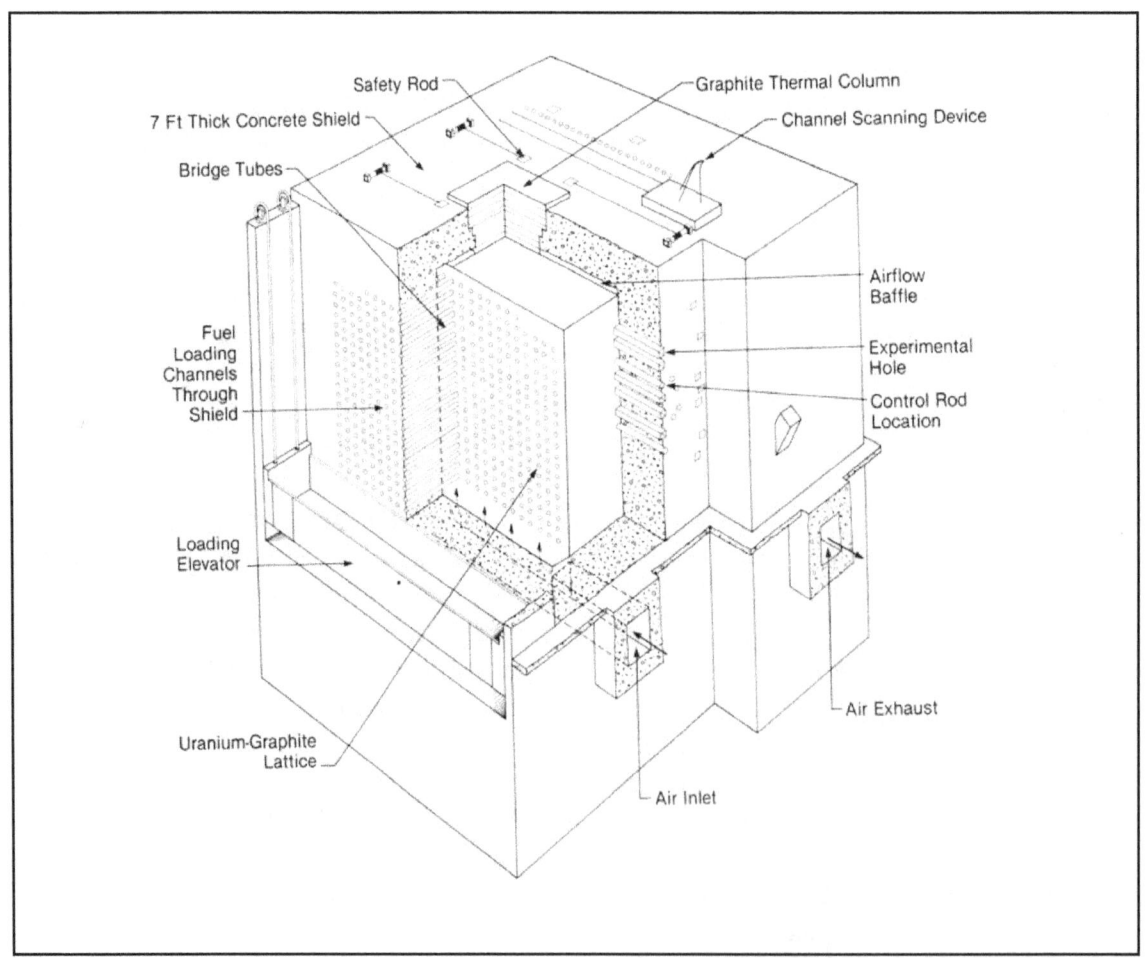

Air-Cooled Pile Built in X-10 Area at Clinton. Reprinted from Richard G. Hewlett and Oscar E. Anderson, Jr., *The New World, 1939-1946*, Volume I of *A History of the United States Atomic Energy Commission* (University Park: Pennsylvania State University Press, 1962).

Workers Loading Uranium Slug Into Face of Air-Cooled Pile. Reprinted from Richard G. Hewlett and Oscar E. Anderson, Jr., *The New World, 1939-1946*, Volume I of *A History of the United States Atomic Energy Commission* (University Park: Pennsylvania State University Press, 1962).

and instrument reliability. Once the production facilities at Oak Ridge and Hanford were underway, however, Met Lab research became increasingly unimportant in the race for the bomb and the scientists found themselves serving primarily as consultants for DuPont.

Decision on Chemical Extraction

While the Met Lab physicists chafed under DuPont domination, a smoother and quieter relationship existed between the chemists and DuPont. Seaborg and Cooper continued to work well together, and enough progress was made in the semiworks for the lanthanum fluoride process in late 1942 that DuPont moved into the plant design stage and converted the semiworks for the bismuth phosphate method. DuPont pressed for a decision on plutonium extraction methods in

late May. Greenewalt chose bismuth phosphate, though even Seaborg admitted he could find little to distinguish between the two. Greenewalt based his decision on the corrosiveness of lanthanum fluoride and on Seaborg's guarantee that he could extract at least fifty percent of the plutonium using bismuth phosphate. DuPont began constructing the chemical separation pilot plant at Oak Ridge, while Seaborg continued refining the bismuth phosphate method.

It was now Cooper's job to design the pile as well as the plutonium extraction facilities at Clinton, both complicated engineering tasks made even more difficult by high levels of radiation produced by the process. Not only did Cooper have to oversee the design and fabrication of parts for yet another new Manhattan Project technology, he had to do so with an eye toward planning the Hanford facility. Safety was a major consideration because of the hazards of working with plutonium, which was highly radioactive. Uranium, a much less active element than plutonium, posed far fewer safety problems.

In July 1942, Compton set up a health division at the Met Lab and put Robert S. Stone in charge. Stone established emission standards and conducted experiments on radiation hazards, providing valuable planning information for the Oak Ridge and Hanford facilities.

Construction at Oak Ridge

DuPont broke ground for the X-10 complex at Oak Ridge in February 1943. The site would include an air-cooled experimental pile, a pilot chemical separation plant, and support facilities. Cooper produced blueprints for the chemical separation plants in time for construction to begin in March. A series of huge underground concrete cells, the first of which sat under the pile, extended to one story above ground. Aluminum cans containing uranium slugs would drop into the first cell of the chemical separation facility and dissolve and then go through the extraction

Oak Ridge

Aerial View of the town of Oak Ridge looking east
In September 1942, General Leslie Groves, U.S. Army Corps of Engineers, designated a 59,000-acre parcel of land between Black Oak Ridge to the north and the Clinch River to the south as the first federal reserve for producing nuclear materials for the atomic bomb. Framed by the foothills of the Appalachian Mountains, the site consisted of family farms and small communities. About 3,000 residents vacated their homes within weeks. One of them later observed: "What do you do? The government needed your land to win the war. Who could refuse such a request as that?"

The first phase of Oak Ridge community planning evolved early in 1943 for a town of around 13,000 residents. This and subsequent planning estimates proved too low. The peak population in September 1945 was 75,000, occupying 10,000 family dwelling units, 13,000 dormitory spaces, 5,000 trailers, and more than 16,000 hutment and barracks accommodations. From 1942 to 1949, Oak Ridge operated as a closed city on the northern edge of the reservation.

Gallaher Ferry
Employees of the Manhattan Engineer District cross the Clinch River on the hand-pulled Gallaher Ferry in June 1943.

K-25 steel-frame construction
The gaseous diffusion plant, designed by the Kellex Corporation, bore the name *K-25*. Because the number 25 was used throughout the project to designate uranium-235, it was added to the title.

Y-12 Plant
As part of the wartime secrecy effort, the designation Y-12 given the electromagnetic plant had no meaning except to confuse enemy agents.

Y-12 Alpha Racetrack under construction

X-10 Pile

Y-12 Alpha Racetrack

Y-12 worker at Beta-3 control panel
Management cautioned male co-workers not to talk to female Y-12 workers as they monitored the control panels.

Worker on lunch break
War workers were drawn from across the nation. Engineers and scientists, clerks and stenographers, technical personnel and general laborers poured into Oak Ridge to work on the project.

War workers commuting by bus
The strict security at Oak Ridge meant that commuters confined their conversations to lighter topics.

Army Post Exchange

Jackson Square
The town's original shopping center.

Lining up for cigarettes at Williams Drug Store, Jackson Square

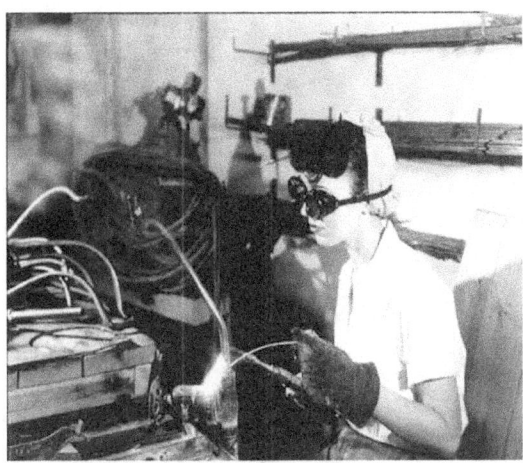

K-25 counterpart of "Rosie-the-Riveter"

X-10 in operation
World's first radioactive isotope produced for medical use, withdrawn from the Graphite Reactor, August 1946.

Y-12 control room
Many women graduates from East Tennessee high schools tended the Y-12 control panels. Each operator monitored two control panels.

Trailer area
Trailers were obtained from the Federal Public Housing Authority. As operating and technical personnel moved into the area, more government trailers were obtained and accommodations provided for privately-owned trailers.

Happy Valley service station
The isolation of the K-25 site forced the Jones Construction Company to build housing facilities for workers. Jones camp, nicknamed "Happy Valley" by its inhabitants, eventually had a total population of about 15,000.

Boom Town
Living conditions were likened to a boom town atmosphere as families struggled to live with uncertain electricity and water supplies. One resident commented: "We weren't born soon enough for the Gold Rush, but we made up for it at Oak Ridge."

Aerial view of trailer park, 1945

East Village area, 1945

Cemesto housing, 1944

The "Jolly Old Elf" encounters tight security at Oak Ridge

Military police at the Elza Gate
Along the perimeter of Oak Ridge stood a settlement known as Elza. Here, the Manhattan Engineer District set up a guarded entrance to control access.

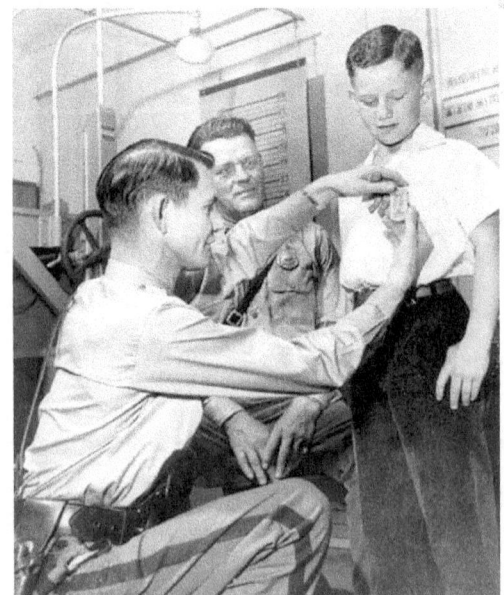

Little boy being badged
Heightened security at Oak Ridge pervaded every aspect of life. Even children wore identification badges.

The "Rhythm Engineers" entertain

Roller skating
The Oak Ridge Recreation and Welfare Association
organized social and recreational activities including
music, art, drama, dancing, and other events like
roller skating.

Teen dance
Dancing was very popular. Segregation laws meant
separate social and recreational events like this
teen dance.

A photograph reveals the veil of secrecy at Oak Ridge
A disabled veteran addresses a bond rally at the K-25 plant and urges the workers to stay on the job.
Army Intelligence cut the negative of this photograph taken December 13, 1944. The left section was published, but the right side, showing the gaseous diffusion plant, remained classified until after the war.

The isolation of East Tennessee made it ideally suited for the secrecy the Manhattan Project required.
One K-25 worker recounted that a military policeman cautioned her that if she ever noticed an odor that she had never smelled before, she and her co-workers should flee immediately. The MP added that he could not tell her more.

Workers and their families accepted the security requirements at Oak Ridge as a routine part of daily living.
"It's all right, son," one mother wrote, "if you can't tell me what you're doing in Oak Ridge, but I do hope it's honorable."

K-25 workers attend to the words of the speaker at the bond rally
The K-25 Plant was a major construction effort. When first built, the building was one of the largest roofed structures in the world, covering nearly 43 acres. Completed at a cost of $500 million and operated by 12,000 workers, the K-25 plant separated uranium-235 from uranium-238.

Construction of the K-25, Y-12, and X-10 plants took place almost simultaneously in 1943. Recruiters were sent throughout the region since there was a shortage of workers. Machinists, plumbers, and other skilled workers found well-paying jobs. For the first time, many East Tennessee women had an opportunity to work outside the home.

"Sunday Punch"
K-25 workers pooled their wages for three Sundays to purchase a North American B-25 Mitchell bomber as a gift to the U.S. Army Air Forces. The workers were hourly workers, or "clockpunchers," thus the name, "Sunday Punch."

Thousands attended the christening of the plane at the Knoxville airport in April 1945. It was assigned to the 12th Bomber Group, 81st Squadron, known as the *Battering Rams.* A member of the squadron, Tom Evans, a 1941 graduate of Knoxville High School (where his father was principal) learned from a friend about the plane. The commander assigned the plane to Evans. Reminiscing years later, Evans said: "I was in my glory. I had my own airplane--and one bought by hometown folks."

Y-12 Workers at Shift Change
This photograph taken in August 1945 vividly demonstrates the role women played at Oak Ridge during the war. Early in May 1943, E.O. Lawrence visited Oak Ridge. He looked down on the Y-12 plant. On his return to Berkeley, he told his colleagues: "Just from the size of the thing, you can see that a thousand people would just be lost in this place, and we've got to make a definite attempt to just hire everybody in sight and somehow use them, because it's going to be an awful job to get those racetracks into operation on schedule."

The racetracks operated around the clock. Women who previously worked as clerks or waitresses found that they could work for the unheard of wage of seventy-five cents an hour. Word of the recruitment effort went out in local papers and friends told friends. Operators worked three shifts over a twenty-four hour period, rotating every seven days. By spring 1945, almost 22,000 workers were required to keep Y-12 in operation.

East Tennesseans celebrate the end of World War II
Relief and joy mingle on the faces of the celebrants at Oak Ridge. At war's end, Oak Ridgers could at last talk about the project.

The first vote in the town of Oak Ridge, August 1946
Early in 1943, the Tennessee Legislature, alarmed by the growing number of Federal projects in the State and the subsequent loss of state lands and taxes from such lands, declined to cede sovereignty to the Federal government over the land taken over by the War Department for the Oak Ridge project. Oak Ridge residents came under the legal jurisdiction of Roane and Anderson Counties as well as the State of Tennessee. After a year's residency, Oak Ridgers could vote in state and county elections.

Hanford
Five hundred sixty square miles of shrub steppe, sand, and sagebrush located on the Columbia River in southeastern Washington State. Selected to serve as the central location for the plutonium plant for the Manhattan Project, the site was given the name Hanford Engineer Works.

Hanford Reach
Of the potential sites investigated, none had a better combination of isolation, abundant water, and surplus hydroelectric power.

Wagon tracks date to pioneer days
The earliest recorded immigration into the area was in 1853, when settlers turned north from the Oregon trail to pass through the Yakima Valley.

Groves at Hanford
General Groves visited Hanford in early 1944 to speak with workers regarding safety issues. Originally the work week consisted of six 8-hour days, but by September 1943 it had expanded to six 9-hour days.

Hanford workers in the HC-11 bank

HC-2 hutments
Both men and women lived in (separate) 16' by 40' hutments in the Hanford Construction Camp. A total of 880 units were built, each housing ten workers.

Aerial view of Richland

Richland
Temporary quarters for construction workers were built in Hanford while permanent facilities for other personnel were located down the road in Richland. Here a neighborhood of mixed housing styles takes shape.

Trailer park at the Hanford townsite

Church
One of five churches built in nearby Richland for Hanford employees and their families.

Library
This library served Hanford employees and their families.

Liberty Theatre at Pasco
The Liberty had a stage for live performances and a full basement for social events. At one end of the basement was a raised platform for the dance bands that often played there. Pasco was the oldest community in the Hanford area.

All the comforts (almost) of home
A Hanford family relaxes on their patio. The pervasive construction excavations slit the fine, ashy soil in so many places that blowing dust and sand buried newly planted lawns, trees, and shrubs.

Shoe repair

Sears Roebuck

Waiting in line at the clothing store

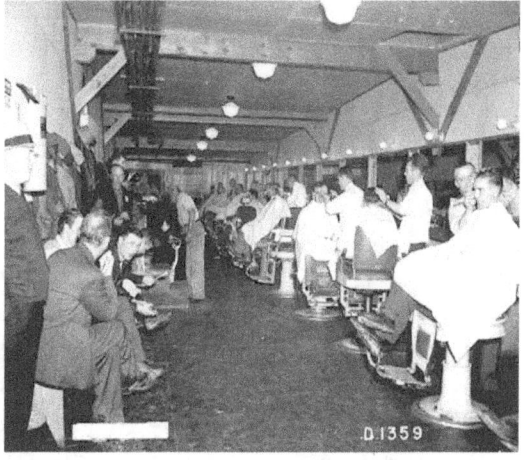

Barber shop
This family owned and operated shop is still in business today.

Sending money home

Batter-up!
America's favorite pastime was taken very seriously by Hanford workers, and the baseball games were a major social occasion.

Game plan

Catcher

The children of Hanford
Hanford had one of the highest birth rates in the nation.
The rapidly expanding families at Manhattan Project sites gave General
Groves concern for the unexpected burden on the medical facilities.
One anonymous poet put it thus:

The General's in a stew
He trusted you and you
He thought you'd be scientific
Instead you're just prolific
And what is he to do?

Santa pays a visit to Hanford

Christmas 1943

Lining up to buy war bonds

Day's Pay
Hanford workers contributed a day's pay to purchase a
Boeing B-17 Flying Fortress that they presented to the
U.S. Army Air Forces in July 1944. The bomber was
appropriately named "Day's Pay."

Conserving for the war effort

T Plant - world's first large-scale plutonium separation facility
The chemical separation buildings, nicknamed "Queen Marys" because of their size and shape, are massive canyon-like structures, 800 feet long, 65 feet wide, and 80 feet high.

T Plant under construction
Facility was completed in December 1944.

T Plant
The white discharge from the stack was sulfur dioxide, a fogging agent used in early meteorological testing at Hanford.

F Reactor looking north
One of three reactors (along with B and D) built about six miles apart on the south bank of the Columbia River.

U Plant
The third process separations facility, U-Plant, was identical to B (East Area) and to T (West Area).

Face of B Reactor under construction
B Reactor was the world's first large scale plutonium production reactor. Built on a much larger scale than the X-10 at Oak Ridge, the B Reactor used water as a coolant.

241-T tank farm under construction, early 1944
High-level radioactive wastes from the chemical separations plants were placed in the finished tanks.

184-B Power House under construction
Provided steam for heating and processing to the B Reactor and surrounding support facilities.

Columbia River Water Pumping Station
Furnished water to the Hanford and White Bluffs living areas.

Manhattan Project
Photo Gallery

Los Alamos

Security Patrol at Los Alamos
General Groves welcomed the isolation offered by Los Alamos—high desert mesa cut by deep canyons into long finger-like extensions, populated by only a few homesteaders.

"The Big House"
Once a dormitory and classroom for the Los Alamos Ranch School, this building housed some of the first scientists to arrive at Los Alamos in 1943.

Homestead cabin near Pajarito Site

Main Hill Road, looking west, leading to Los Alamos
Early residents described the drive to Los Alamos as nerve-wracking.

Highway to the "secret city"
Road construction on State Highway 4, one of two alternative routes into Los Alamos, was begun in September 1943.

A bumpy ride
Winter snows and summer rains turned roads into muddy tracks.

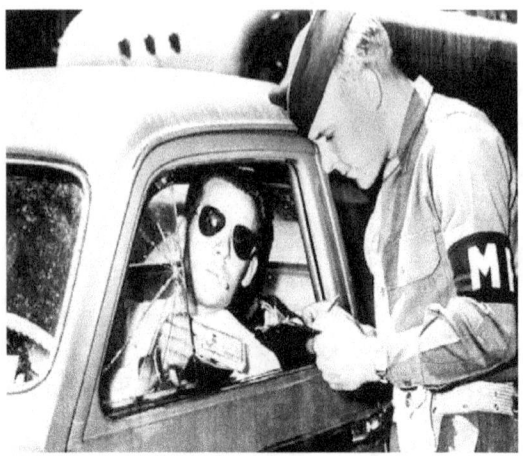

A resident shows his pass at the Main Gate
Life as you knew it ended at Project Y's Main Gate
where your identity was exchanged for a number.

Technical Area 1
Guards stationed at the Main Gate checked cars and
passengers for proper identification. Technical area
personnel went through another set of gates to enter
the laboratory.

Los Alamos parking space

Home sweet home
New arrivals to Project Y were struck by the eclectic housing mix: rows of barracks, apartments, Quonset huts, government trailers, and prefabricated units lined row upon row of nameless, unpaved streets.

Wartime general store

Kitchen and dinette in a Morgan housing unit
Houses were supplied with modern refrigerators but had antiquated stoves, nicknamed Black Beauties. Many residents gave up trying to learn how to operate these "beauties" and used hotplates instead.

Main Cafeteria, Fuller Lodge, 1946

Hanford-style housing unit, used as self-help laundry, 1945

Laundry day
Water, or the lack of it, was a constant problem. Soldiers hand delivered bulletins to Los Alamos residents with precise instructions on water use.

Slotin criticality accident, Pajarito site
Within a year of the Trinity test, Louis Slotin perished
when a plutonium core started an unintended fission
chain reaction.

Ernest Lawrence, Enrico Fermi, and Isidor Rabi, 1945

Handle with care
Pulling a radioactive source from a storage building. Careful handling of large radioactive sources was
accomplished by crude methods in the early days. There were a number of female technicians at Los Alamos.

Trinity base camp
The base camp included barracks, a mess hall, warehouses, and an explosive magazine.
At the time of the test, there were over 200 camp residents.

The road to Trinity
Military Police examine special credentials at an intersection designated Broadway and Pennsylvania Avenues at the Trinity test site.

McDonald ranch house, Trinity Site
The McDonald ranch house was used for the final assembly of the vital components.

Jumbo's desert journey
A specially designed 64-wheel trailer began the journey of carrying Jumbo across the desert from a nearby railroad siding to the Trinity Site. Originally intended to be a part of the Trinity test, the 200-ton container was eliminated during final planning.

South Yard Bunker
Three observation bunkers located 10,000 yards north, west, and south of the Trinity Tower were used to determine the symmetry of the implosion and the amount of energy released.

Valued Cargo
Vital components for the device are brought into the McDonald ranch house in a shock-proof case.

The Trinity fireball, 0.053 seconds after detonation, as it shook the desert near the town of San Antonio, New Mexico, on July 16, 1945

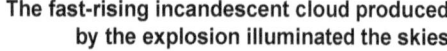

The fast-rising incandescent cloud produced by the explosion illuminated the skies

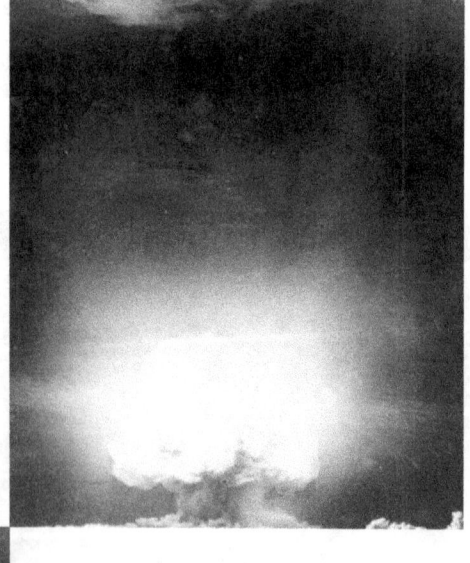

15 seconds into detonation, the Trinity fireball rises into the air above the desert "The whole country was lighted by a searing light with the intensity many times that of the midday sun," recalled General Thomas Farrell.

J. Robert Oppenheimer and General Leslie Groves and the remains of the Trinity Test Tower, September 1945

The Tower Marker at Trinity

At war's end, the Laboratory received the Army/Navy E (for excellence) award

Aerial View of Hanford Community. Reprinted from Richard G. Hewlett and Oscar E. Anderson, Jr., *The New World, 1939-1946,* Volume I of *A History of the United States Atomic Energy Commission* (University Park: Pennsylvania State University Press, 1962).

process. The pile building went up during the spring and summer, a huge concrete shell seven feet thick with hundreds of holes for uranium slug placement. Slugs were to plutonium piles what barrier was to gaseous diffusion; that is, an obstacle that could shut down the entire process. The Aluminum Company of America (Alcoa) was the only firm left working on a process to enclose uranium-235 in aluminum sheaths, and it was still having problems. Initial production provided mixed results, with many cans failing vacuum tests because of faulty welds.

X-10 in Operation: Fall 1943

The moment everyone had been waiting for came in late October when DuPont completed construction and tests of the X-10 pile at Clinton Engineer Works. After thousands of slugs were loaded, the pile went critical in the early morning of November 4 and produced plutonium by the end of the month. Criticality was achieved with only half of the channels filled with uranium. During the next several months, Compton gradually raised the power level of the pile and increased its plutonium yield. Chemical separation techniques using the bismuth phosphate process were so successful that Los Alamos received plutonium samples beginning in the spring. Fission studies of these samples at Los Alamos during summer 1944 heavily influenced bomb design.

Pile D at Hanford. Pile in Foreground, Water Treatment Plant in Rear. Reprinted from Richard G. Hewlett and Oscar E. Anderson, Jr., *The New World, 1939-1946,* Volume I of *A History of the United States Atomic Energy Commission* (University Park: Pennsylvania State University Press, 1962).

Hanford Takes Shape

Colonel Matthias returned to the Hanford area to set up a temporary office on February 22, 1943. His orders were to purchase half a million acres in and around the Hanford-Pasco-White Bluffs area, a sparsely populated region where sheep ranching and farming were the main economic activities. Many of the area's landowners rejected initial offers on their land and took the Army to court seeking more acceptable appraisals. Matthias adopted a strategy of settling out of court to save time, time being a more important commodity than money to the Manhattan Project.

Matthias received his assignment in late March. The three water-cooled piles, designated by the letters B, D, and F, would be built about six miles apart on the south bank of the Columbia River. The four chemical separation plants, built in pairs, would be nearly ten miles south of the piles, while a facility to produce slugs and perform tests would be approximately twenty miles southeast of the separation plants near Richland. Temporary quarters for construction workers would be put up in Hanford, while permanent facilities for other personnel would be located down the road in Richland, safely removed from the production and separation plants.

During summer 1943, Hanford became the Manhattan Project's newest atomic boomtown. Thousands of workers poured into the town, many of them to leave in discontent. Well situated from a logistical point of view, Hanford was a sea of tents and barracks where workers had little to do and nowhere to go. DuPont and the Army coordinated efforts to recruit laborers from all over the country for Hanford, but even with a relative labor surplus in the Pacific Northwest, shortages plagued the project. Conditions improved significantly during the second half of the year, with the addition of recreational facilities, higher pay, and better overall services for Hanford's population, which reached 50,000 by summer 1944. Hanford still resembled the frontier and mining towns once common in the west, but the rate of worker turnover dropped substantially.

Groundbreaking for the water-cooling plant for the 100-B pile, the westernmost of the three, took place on August 27, less than two weeks before Italy's surrender to the Allies on September 8. Work on the pile itself began in February, with the base and shield being completed by mid-May. It took another month to place the graphite pile

Chemical Separation Plant (Queen Mary) Under Construction at Hanford. Reprinted from Richard G. Hewlett and Oscar E. Anderson, Jr., *The New World, 1939-1946,* Volume I of *A History of the United States Atomic Energy Commission* (University Park: Pennsylvania State University Press, 1962).

and install the top shield and two more months to wire and pipe the pile and connect it to the various monitoring and control devices.

At Hanford, irradiated uranium slugs would drop into water pools behind the piles and then be moved by remote-controlled rail cars to a storage facility five miles away for transportation to their final destination at one of the two chemical separation locations, designated 200-West and 200-East. The T and U plants were located at 200-West, while a single plant, the B unit, made up the 200-East complex (the planned fourth chemical separation plant was not built). The Hanford chemical separation facilities were massive scaled-up versions of those at Oak Ridge, each containing separation and concentration buildings in addition to ventilation (to eliminate radioactive and poisonous gases) and waste storage areas. Labor shortages and the lack of final blueprints forced DuPont to stop work on the 200 areas in summer 1943 and concentrate its forces on 100-B, with the result that 1943 construction progress on chemical separation was limited to digging two huge holes in the ground.[37]

The Chemical Separation Buildings (Queen Marys)

Both 221T and 221U, the chemical separation buildings in the 200-West complex, were finished by December 1944. 221B, their counterpart in 200-East, was completed in spring 1945. Nicknamed Queen Marys by the workers who built them, the separation buildings were awesome canyon-like structures 800 feet long, 65 feet wide, and 80 feet high containing forty process pools. The interior had an eerie quality as operators behind thick concrete shielding manipulated remote control equipment by looking through television monitors and periscopes from an upper gallery. Even with massive concrete lids on the process pools, precautions against radiation exposure were necessary and influenced all aspects of plant design.[38]

Completed Queen Mary at Hanford. Reprinted from Richard G. Hewlett and Oscar E. Anderson, Jr., *The New World, 1939-1946*, Volume I of *A History of the United States Atomic Energy Commission* (University Park: Pennsylvania State University Press, 1962).

Construction of the chemical concentration buildings (224-T, -U, and -B) was a less daunting task because relatively little radioactivity was involved, and the work was not started until very late 1944. The 200-West units were finished in early October, the East unit in February 1945. In the Queen Marys, bismuth phosphate carried the plutonium through the long succession of process pools. The concentration stage was designed to separate the two chemicals. The normal relationship between pilot plant and production plant was realized when the Oak Ridge pilot plant reported that bismuth phosphate was not suitable for the concentration process but that Seaborg's original choice, lanthanum fluoride, worked quite well. Hanford, accordingly, incorporated this suggestion into the concentration facilities. The final step in plutonium extraction was isolation, performed in a more typical laboratory setting with little radiation present. Here Perlman's earlier research on the peroxide method paid off and was applied to produce pure plutonium nitrate.

The nitrate would be converted to metal in Los Alamos, New Mexico.

Los Alamos

The final link in the Manhattan Project's farflung network was the Los Alamos Scientific Laboratory in Los Alamos, New Mexico. The laboratory that designed and fabricated the first atomic bombs, codenamed Project Y, began to take shape in spring 1942 when Conant suggested to Bush that the Office of Scientific and Research Development and the Army form a committee to study bomb development. Bush agreed and forwarded the recommendation to Vice President Wallace, Secretary of War Stimson, and General Marshall (the Top Policy Group). By the time of his appointment in late September, Groves had orders to set up a committee to study military applications of the bomb. Meanwhile, sentiment was growing among the Manhattan scientists that research on the bomb project needed to be better coordinated. Oppenheimer, among others, advocated a central facility where theoretical and experimental work could be conducted according to standard scientific protocols. This would insure accuracy and speed progress. Oppenheimer suggested that the bomb laboratory operate secretly in an isolated area but allow free exchange of ideas among the scientists on the staff. Groves

accepted Oppenheimer's suggestion and began seeking an appropriate location.

The search for a bomb laboratory site quickly narrowed to two places in northern New Mexico, Jemez Springs and the Los Alamos Boys Ranch School, locations Oppenheimer knew well since

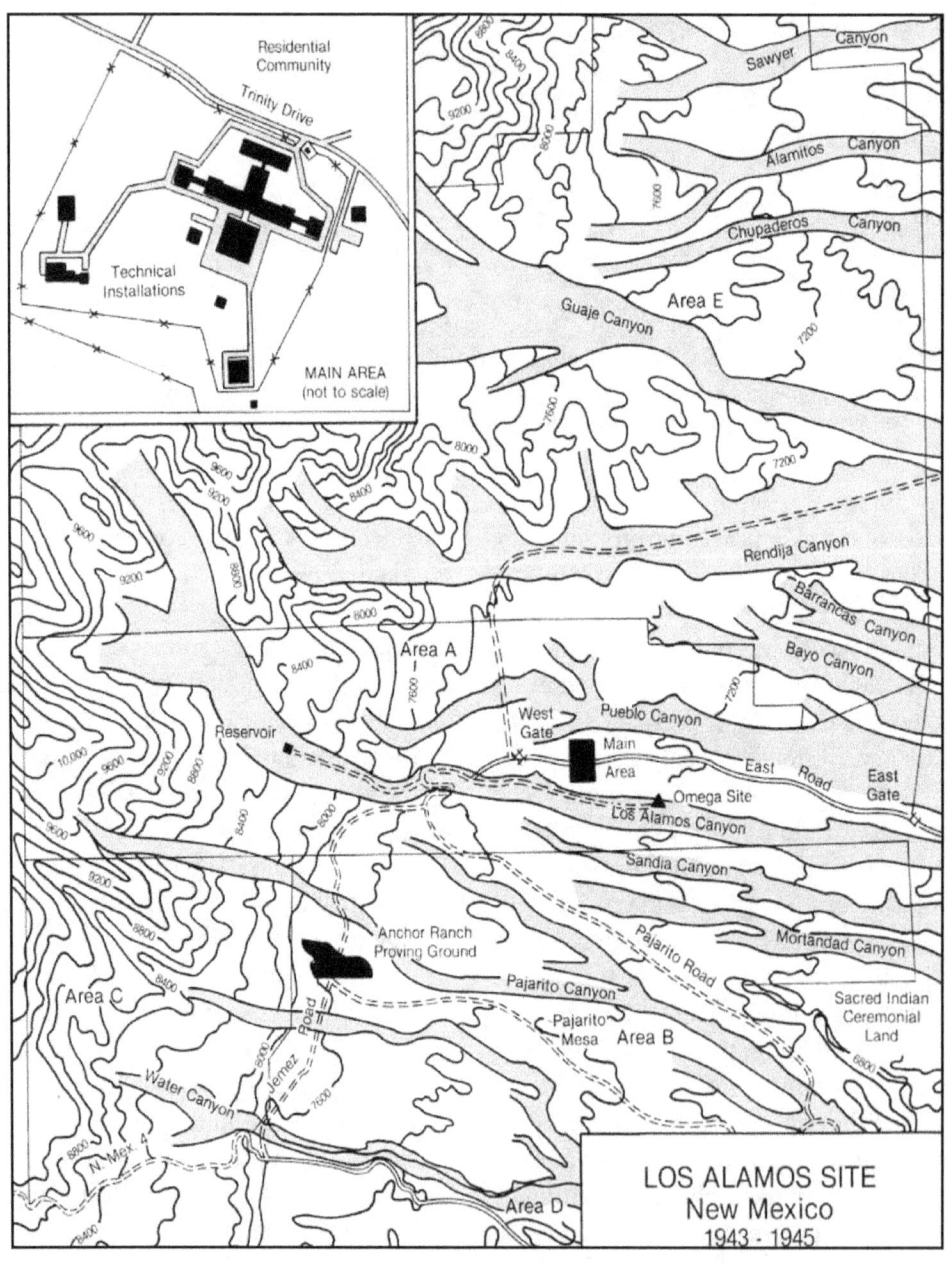

Los Alamos Site. Reprinted from Vincent C. Jones, *Manhattan: The Army and the Atomic Bomb* (Washington, D.C.: U.S. Government Printing Office, 1985).

he had a ranch nearby in the Pecos Valley of the Sangre de Cristo Mountains. In mid-November, Oppenheimer, Groves, Edwin M. McMillan, and Lieutenant Colonel W. H. Dudley visited the two sites and chose Los Alamos. Located on a mesa about thirty miles northwest of Santa Fe, Los Alamos was virtually inaccessible. It would have to be provided with better water and power facilities, but the laboratory community was not expected to be very large. The boys' school occupying the site was eager to sell, and Groves was equally eager to buy. By the end of 1942 the district engineer in Albuquerque had orders to begin construction, and the University of California had contracted to provide supplies and personnel.

Oppenheimer and Groves

Oppenheimer, selected to head the new laboratory, proved to be an excellent director despite initial concerns about his administrative inexperience, leftist political sympathies, and lack of a Nobel Prize when several scientists he would be directing were prizewinners. Groves worked well with Oppenheimer although the two were fundamentally different in temperament. Groves was a practical-minded military man, brusque and goal-oriented. His aide, Colonel Nichols, characterized his heavyset boss as ruthless, egotistical, and confident, "the biggest S.O.B. I have ever worked for. He is most demanding. He is most critical. He is always a driver, never a praiser. He is abrasive and sarcastic." Nichols admitted, however, that if he had it to do over again, he would once again "pick General Groves [as his boss]" because of his unquestioned ability.[39] Groves demanded that the Manhattan Project scientists spend all their time on the bomb and resist the temptation, harmless enough in peacetime, to follow lines of research that had no direct applicability to immediate problems. In contrast to Groves, Oppenheimer was a philosophical man, attracted to Eastern mysticism and of a decidedly theoretical inclination and

Groves and Oppenheimer. Reprinted from Leslie R. Groves, *Now It Can Be Told* (New York: Harper & Row, 1962).

sensitive nature. A chainsmoker given to long working hours, Oppenheimer appeared almost emaciated. The Groves-Oppenheimer alliance, though not one of intimacy, was marked by mutual respect and was a major factor in the success of the Manhattan Project.

Oppenheimer insisted, with some success, that scientists at Los Alamos remain as much an academic community as possible, and he proved adept at satisfying the emotional and intellectual needs of his highly distinguished staff. Hans Bethe, head of the theoretical division, remembered that nobody else in that laboratory "even came close to him. In his knowledge. There was human warmth as well. Everybody certainly had the impression that Oppenheimer cared what each particular person was doing. In talking to someone he made it clear that that person's work was important for the success of the whole project."[40]

Oppenheimer had a chance to display his persuasive abilities early when he had to convince scientists, many of them already deeply involved in war-related research in university laboratories, to join his new organization. Complicating his task were the early plans to operate Los Alamos as a military laboratory. Oppenheimer accepted Groves's rationale for this arrangement but soon found that scientists objected to working as commissioned officers and feared that the military chain of command was ill-suited to scientific decision making. The issue came to a head when Oppenheimer tried to convince Robert F. Bacher and Isidor I. Rabi of the Massachusetts Institute of Technology's Radiation Laboratory to join the Los Alamos team. Neither thought a military environment was conducive to scientific research. At Oppenheimer's request, Conant and Groves wrote a letter explaining that the secret weapon-related research had presidential authority and was of the utmost national importance. The letter promised that the laboratory would remain civilian through 1943, when it was believed that security would require militarization of the final stages of the project (in fact, militarization never took place). Oppenheimer would supervise all scientific work, and the military would maintain the post and provide security.

Recruiting the Staff

Oppenheimer spent the first three months of 1943 tirelessly crisscrossing the country in an attempt to put together a first-rate staff, an effort that proved highly successful.[41] Even Bacher signed on, though he promised to resign the moment militarization occurred; Rabi, though he did not move to Los Alamos, became a valuable consultant. As soon as Oppenheimer arrived at Los Alamos in mid-March, recruits began arriving from universities across the United States, including California, Minnesota, Chicago, Princeton, Stanford, Purdue, Columbia, Iowa State, and the Massachusetts Institute of Technology, while still others came from the Met Lab and the National Bureau of Standards.

Virtually overnight Los Alamos became an ivory tower frontier boomtown, as scientists and their families, along with nuclear physics equipment, including two Van de Graaffs, a Cockroft-Walton accelerator, and a cyclotron, arrived caravan fashion at the Santa Fe railroad station and then made their way up to the mesa along the single primitive road. It was a most remarkable collection of talent and machinery that settled this remote outpost of the Manhattan Project.

Theory and the "Gadget"

The initial spartan environment of "the Hill" (which included box lunches and temporary housing) was without doubt quite a contrast to the comfortable campus settings so familiar to many on the staff. But the laboratory's work began even as the Corps of Engineers struggled to provide the amenities of civilized life. The properties of uranium were reasonably well understood, those of plutonium less so, and knowledge of fission explosions entirely theoretical. That 2.2 secondary neutrons were produced when uranium-235 fissioned was accepted, but while Seaborg's team had proven in March 1941 that plutonium underwent neutron-induced fission, it was not known yet if plutonium released secondary neutrons during bombardment. The theoretical consensus was that chain reactions took place with sufficient speed to produce powerful releases of energy and not simply explosions of the critical mass itself, but only experiments could test the theory. The optimum size of the critical mass remained to be established, as did the optimum shape. When enough data were gathered to establish optimum critical mass, optimum effective mass still had to be determined. That is, it was not enough simply to start a chain reaction in a critical mass; it was necessary to start one in a mass that would release the greatest possible amount of energy before it was destroyed in the explosion.

In addition to calculations on uranium and plutonium fission, chain reactions, and critical and effective masses, work needed to be done on the ordnance aspects of the bomb, or "gadget" as it came to be known. Two subcritical masses of fissionable material would have to come together to form a supercritical mass for an explosion to occur. Furthermore, they had to come together in a precise manner and at high speed. Measures also had to be taken to insure that the highly unstable subcritical masses did not predetonate because of spontaneously emitted neutrons or neutrons produced by alpha particles reacting with lightweight impurities. The chances of predetonation could be reduced by purification of the fissionable material and by using a high-speed firing system capable of achieving velocities of 3,000 feet per second. A conventional artillery method of firing one subcritical mass into the other was under consideration for uranium-235, but this method would work for plutonium only if absolute purification of plutonium could be achieved.

A variation of the artillery method was designed for uranium. Bomb designers, unable to solve the purification problem, turned to the relatively unknown implosion method for plutonium. With implosion, symmetrical shockwaves directed inward would compress a subcritical mass of plutonium packed in a nickel casing (tamper), releasing neutrons and causing a chain reaction.

Always in the background loomed the hydrogen bomb, a thermonuclear device considerably more powerful than either a uranium or plutonium device but one that needed a nuclear fission bomb as a detonator. Research on the hydrogen bomb, or Super, was always a distant second in priority at Los Alamos, but Oppenheimer concluded that it was too important to ignore. After considerable thought, he gave Teller permission to devote himself to the Super. To make up for Teller's absence, Rudolf Peierls, one of a group of British scientists who reinforced the Los Alamos staff at the beginning of 1944, was added to Bethe's theory group in mid-1944. Another member of

the British contingent was the Soviet agent Klaus Fuchs, who had been passing nuclear information to the Russians since 1942 and continued doing so until 1949 when he was caught and convicted of espionage (and subsequently exchanged).[42]

Another Lewis Committee

The first few months at Los Alamos were occupied with briefings on nuclear physics for the technical staff and with planning research priorities and organizing the laboratory. Groves called once again on Warren Lewis to head a committee, this time to evaluate the Los Alamos program. The committee's recommendations resulted in the coordinated effort envisioned by those who advocated a unified laboratory for bomb research. Fermi took control of critical mass experiments and standardization of measurement techniques. Plutonium purification work, begun at the Met Lab, became high priority at Los Alamos, and increased attention was paid to metallurgy. The committee also recommended that an engineering division be organized to collaborate with physicists on bomb design and fabrication. The laboratory was thus organized into four divisions: theoretical (Hans A. Bethe); experimental physics (Robert F. Bacher); chemistry and metallurgy (Joseph W. Kennedy); and ordnance (Navy Captain William S. "Deke" Parsons). Like other Manhattan Project installations, Los Alamos soon began to expand beyond initial expectations.

As director, Oppenheimer shouldered burdens both large and small, including numerous mundane matters such as living quarters, mail censorship, salaries, promotions, and other "quality of life" issues inevitable in an intellectual pressure-cooker with few social amenities. Oppenheimer relied on a group of advisers to help him keep the "big picture" in focus, while a committee made up of Los Alamos group leaders provided day-to-day communications between divisions.

Early Progress

Early experiments on both uranium and plutonium provided welcome results. Uranium emitted neutrons in less than a billionth of a second—just enough time, in the world of nuclear physics, for an efficient explosion. Emilio Segrè later provided an additional cushion with his discovery in December 1943 that, if cosmic rays were eliminated, the subcritical uranium masses would not have to be brought together as quickly as previously thought; nor would the uranium have to be as pure. Muzzle velocity for the scaled-down artillery piece could be lower, and the gun could be shorter and lighter.[43] Segrè's tests on the first samples of plutonium demonstrated that plutonium emitted even more neutrons than uranium due to the spontaneous fission of plutonium-240. Both theory and experimental data now agreed that a bomb using either element would detonate if it could be designed and fabricated into the correct size and shape. But many details remained to be worked out, including calculations to determine how much uranium-235 or plutonium would be needed for an explosive device.

Bacher's engineering division patiently generated the essential cross-sectional measurements needed to calculate critical and efficient mass. (The cross section is a measurement that indicates the probability of a nuclear reaction taking place.) The same group utilized particle accelerators to produce the large numbers of neutrons needed for its cross-sectional experiments. Bacher's group also compiled data that helped identify tamper materials that would most effectively push neutrons back to the core and enhance the efficiency of the explosion. Despite Los Alamos's postwar reputation as a mysterious retreat where brilliant scientists performed miracles of nuclear physics, much of the work that led to the atomic bombs was extremely tedious.

The chemists' job was to purify the uranium-235 and plutonium, reduce them to metals, and process the tamper material. Only highly purified uranium and plutonium would be safe from predetonation. Fortunately purification standards for uranium were relatively modest, and the chemical division was able to focus its effort on the lesser known plutonium and make substantial progress on a multi-step precipitation process by summer 1944. The metallurgy division had to turn the purified uranium-235 and plutonium into metal. Here, too, significant progress was made by summer as the metallurgists adapted a stationary-bomb technique initially developed at Iowa State University.

Parsons, in charge of ordnance engineering, directed his staff to design two artillery pieces of relatively standard specifications except for their extremely light barrels—one for a uranium weapon and one for a plutonium bomb. The weapons needed to achieve high velocities, but they would not have to be durable since they would only be fired once. Here again early efforts centered on the more problematic plutonium weapon, which required a higher velocity due to its higher risk of predetonation. Two plutonium guns arrived in March and were field-tested successfully. In the same month, two uranium guns were ordered.

Early Implosion Work

Parsons assigned implosion studies a low priority and placed the emphasis on the more familiar artillery method. Consequently, Seth H. Neddermeyer performed his early implosion tests in relative obscurity. Neddemeyer found it difficult to achieve symmetrical implosions at the low velocities he had achieved. When the Princeton mathematician John von Neumann, a Hungarian refugee, visited Los Alamos late in 1943, he suggested that high-speed assembly and high velocities would prevent predetonation and achieve more symmetrical explosions. A relatively small, subcritical mass could be placed under so much pressure by a symmetrical implosion that an efficient detonation would occur. Less critical material would be required, bombs could be ready earlier, and extreme purification of plutonium would be unnecessary. Von Neumann's theories excited Oppenheimer, who assigned Parsons's deputy, George B. Kistiakowsky, the task of perfecting implosion techniques. Because Parsons and Neddemeyer did not get along, it was Kistiakowsky who worked with the scientists on the implosion project. While experiments on implosion and explosion continued, Parsons directed much of his effort toward developing bomb hardware, including arming and wiring mechanisms and fuzing devices. Working with the Army Air Force, Parsons's group developed two bomb models by March 1944 and began testing them with B-29s. Thin Man, named for President Roosevelt, utilized the plutonium gun design, while Fat Man, named after Winston Churchill, was an implosion prototype. (Segrè's lighter, smaller uranium gadget became Little Boy, Thin Man's brother.)

Elimination of Thin Man

Thin Man was eliminated four months later because of the plutonium-240 contamination problem. Seaborg had warned that when plutonium-239 was irradiated for a length of time it was likely to pick up an additional neutron, transforming it into plutonium-240 and increasing the danger of predetonation (the bullet and target in the plutonium weapon would melt before coming together). Measurements taken at Clinton confirmed the presence of plutonium-240 in the plutonium produced in the experimental pile. On July 17 the difficult decision was made to cease work on the plutonium gun method. Plutonium could be used only in an implosion device, but in summer 1944 an implosion weapon looked like a long shot.

Abandonment of the plutonium gun project eliminated a shortcut to the bomb. This necessitated a revision of the estimates of weapon delivery Bush had given the President in 1943.

The new timetable, presented to General Marshall by Groves on August 7, 1944—two months after the Allied invasion of France began at Normandy on June 6—promised small implosion weapons of uranium or plutonium in the second quarter of 1945 if experiments proved satisfactory. More certain was the delivery of a uranium gun bomb by August 1, 1945, and the delivery of one or two more by the end of that year. Marshall and Groves acknowledged that German surrender might take place by summer 1945, thus making it probable that Japan would be the target of any atomic bombs ready at that time.

Question Marks: Summer 1944

It was still unclear if even the August 1 deadline could be met. While expenditures reached

Section of S-50 Liquid Thermal Diffusion Plant at Clinton. Reprinted from Richard G. Hewlett and Oscar E. Anderson, Jr., *The New World, 1939-1946,* Volume I of *A History of the United States Atomic Energy Commission* (University Park: Pennsylvania State University Press, 1962).

$100 million per month by mid-1944, the Manhattan Project's goal of producing weapons for the current war was not assured. Operational problems plagued the Y-12 electromagnetic facility just coming on line. The K-25 gaseous diffusion plant threatened to become an expensive white elephant if suitable barriers could not be fabricated. And the Hanford piles and separation facilities faced an equally serious threat as not enough of the uranium-containing slugs to feed the pile were available. Even assuming that enough uranium or plutonium could be delivered by the production facilities built in such great haste, there was no guarantee that the Los Alamos laboratory would be able to design and fabricate weapons in time. Only the most optimistic in the Manhattan Project would have predicted, as Groves did when he met with Marshall, that a bomb or bombs powerful enough to make a difference in the current war would be ready by August 1, 1945.

Progress at Oak Ridge

During winter 1944-45, there was substantial progress at Oak Ridge, thanks to improved performance in each of the production facilities and Nichols's work in coordinating a complicated feed schedule that maximized output of enriched uranium by utilizing the electromagnetic, thermal diffusion, and gaseous diffusion processes in tandem. Nine Alpha and three Beta racetracks were operational and, while not producing up to design potential, were becoming significantly more reliable because of maintenance improvements and chemical refinements introduced by Tennessee Eastman. The S-50 thermal diffusion plant being built by the H. K. Ferguson Company was almost complete and was producing small amounts of enriched material in the finished racks, and the K-25 gaseous diffusion plant, complete with barriers, was undergoing final leak tests. By March 1945, Union Carbide had worked out most of the kinks in K-25 and had started recycling uranium hexafluoride through the system. S-50 was finished at the same time

that the Y-12 racetracks were demonstrating increased efficiency. The Beta calutrons at the electromagnetic plant were producing weapon-grade uranium-235 using feed from the modified Alpha racetracks and the small output from the gaseous diffusion and thermal diffusion facilities. Oak Ridge was now sending enough enriched uranium-235 to Los Alamos to meet experimental needs. To increase production, Groves proposed an additional gaseous diffusion plant (K-27) for low-level enrichment and a fourth Beta track for high-level enrichment, both to be completed by February 1946, in time to contribute to the war against Japan, which many thought would not conclude before summer 1946.

Hanford's Role

With the abandonment of the plutonium gun bomb in July 1944, planning at Hanford became more complicated. Pile 100-B was almost complete, as was the first chemical separation plant, while pile D was at the halfway point. Pile F was not yet under construction. If implosion devices using plutonium could be developed at Los Alamos, the three piles would probably produce enough plutonium for the weapons required, but as yet no one was sure of the amount needed.

Pile Operation

Excitement mounted at Hanford as the date for pile start-up approached. Fermi placed the first slug in pile 100-B on September 13, 1944. Final checks on the pile had been uneventful. The scientists could only hope they were accurate, since once the pile was operational the intense radioactivity would make maintenance of many components impossible. Loading slugs and taking measurements took two weeks. From just after midnight until approximately 3:00 a.m. on September 27, the pile ran without incident at a power level higher than any previous chain reaction (though only at a fraction of design capacity). The operators were elated, but their excitement turned to astonishment when the power level began falling after three hours. It fell continuously until the pile ceased operating entirely on the evening of the 28th. By the next morning the reaction began again, reached the previous day's level, then dropped.

Xenon Poisoning

Hanford scientists were at a loss to explain the pile's failure to maintain a chain reaction. Only the foresight of DuPont's engineers made it possible to resolve the crisis. The cause of the strange phenomenon proved to be xenon poisoning. Xenon, a fission product isotope with a mass of 135, was produced as the pile operated. It captured neutrons faster than the pile could produce them, causing a gradual shutdown. With shutdown, the xenon decayed, neutron flow began, and the pile started up again. Fortuitously, despite the objections of some scientists who complained of DuPont's excessive caution, the company had installed a large number of extra tubes. This design feature meant that pile 100-B could be expanded to reach a power level sufficient to overwhelm the xenon poisoning. Success was achieved when the first irradiated slugs were discharged from pile 100-B on Christmas Day, 1944. The irradiated slugs, after several weeks of storage, went to the chemical separation and concentration facilities. By the end of January 1945, the highly purified plutonium underwent further concentration in the completed chemical isolation building, where remaining impurities were removed successfully. Los Alamos received its first plutonium on February 2.[44]

Reorganizing for the Final Push

Oppenheimer acted quickly to maximize the laboratory's efforts to master implosion. Only if the implosion method could be perfected would the plutonium produced at Hanford come into play. Without either a plutonium gun bomb or implosion weapon, the burden would fall entirely on uranium and the less efficient gun method. Oppenheimer directed a major reorganization of Los Alamos in July 1944 that prepared the way

for the final development of an implosion bomb. Robert Bacher took over G Division (for gadget) to experiment with implosion and design a bomb; George Kistiakowsky led X Division (for explosives) in work on the explosive components; Hans Bethe continued to head up theoretical studies; and "Deke" Parsons now focused on overall bomb construction and delivery.

Field tests performed with uranium-235 prototypes in late 1944 eased doubts about the artillery method to be employed in the uranium bomb. It was clear that the uranium-235 from Oak Ridge would be used in a gun-type nuclear device to meet the August 1 deadline Groves had given General Marshall and the Joint Chiefs of Staff. The plutonium produced at such expense and effort at Hanford would not fit into wartime planning unless a breakthrough in implosion technology occurred.

At the same time, Los Alamos shifted from research to development and production. Time was of the essence, though laboratory research had not yet charted a clear path to the final product. Army Air Force training could wait no longer, and in September at Wendover Field in western Utah, Colonel Paul Tibbets began drilling the 393rd Bombardment Squadron, the heart of the 509th Composite Wing, in test drops with 5,500-pound orange dummy bombs, nicknamed pumpkins. In June 1945, Tibbets and his command moved to Tinian Island in the Marianas, where the Navy SeaBees had built the world's largest airport to accommodate Boeing's new B-29 Superfortresses.

Taking Care of Business

Personnel shortages, particularly of physicists, and supply problems complicated Oppenheimer's task. The procurement system, designed to protect the secrecy of the Los Alamos project, led to frustrating delays and, when combined with persistent late war shortages, proved a constant headache. The lack of contact between the remote laboratory and its supply sources exacerbated the problem, as did the relative lack of experience the academic scientists had with logistical matters.

Groves and Conant were determined not to let mundane problems compromise the bomb effort, and in fall 1944 they made several changes to prevent this possibility. Conant shipped as many scientists as could be spared from Chicago and Oak Ridge to Los Alamos, hired every civilian machinist he could lay his hands on, and arranged for Army enlisted men to supplement the work force (these GIs were known as SEDs, for Special Engineering Detachment). Hartley Rowe, an experienced industrial engineer, provided help in easing the transition from research to production. Los Alamos also arranged for a rocket research team at the California Institute of Technology to aid in procurement, test fuzes, and contribute to component development. These changes kept Los Alamos on track as weapon design reached its final stages.

Freezing Weapon Design

Weapon design for the uranium gun bomb was frozen in February 1945. Confidence in the weapon was high enough that a test prior to combat use was seen as unnecessary. The design for an implosion device was approved in March with a test of the more problematic plutonium weapon scheduled for July 4. Oppenheimer shifted the laboratory into high gear and assigned Allison, Bacher, and Kistiakowsky to the Cowpuncher Committee to "ride herd" on the implosion weapon. He placed Kenneth T. Bain-

bridge in charge of Project Trinity, a new division to oversee the July test firing. Parsons headed Project Alberta, known as Project A, which had the responsibility for preparing and delivering weapons for combat.

During these critical months, much depended upon the ability of the chemists and metallurgists to process the uranium and plutonium into metals and craft them into the correct shape and size. Plutonium posed by far the greater obstacle. It existed in different states, depending upon temperature, and was extremely toxic. Working under intense pressure, the chemists and metallurgists managed to develop precise techniques for processing plutonium just before it arrived in quantity beginning in May.

As a result of progress at Oak Ridge and metallurgical and chemical refinements on plutonium that improved implosion's chances, the nine months between July 1944 and April 1945 saw the American bomb project progress from doubtful to probable. The August 1 delivery date for the Little Boy uranium bomb certainly appeared more likely than it had when Groves briefed Marshall. There would be no implosion weapons in the first half of 1945 as Groves had hoped, but developments in April boded well for the scheduled summer test of the Fat Man plutonium bomb. And recent calculations provided by Bethe's theoretical group gave hope that the yield for the first weapon would be in the vicinity of 5,000 tons of TNT rather than the 1,000-ton estimate provided in fall 1944.

Part V:
The Atomic Bomb and American Strategy

With the Manhattan Project on the brink of success in spring 1945, the atomic bomb became an increasingly important element in American strategy. A long hoped-for weapon now seemed within reach at a time when hard decisions were being made, not only on ending the war in the Pacific, but also on the shape of the postwar international order.

From Roosevelt to Truman

On April 12, only weeks before Germany's unconditional surrender on May 7, President Roosevelt died suddenly in Warm Springs, Georgia, bringing Vice President Harry S. Truman, a veteran of the United States Senate, to the presidency. Truman was not privy to many of the secret war efforts Roosevelt had undertaken and had to be briefed extensively in his first weeks in office. One of these briefings, provided by Secretary of War Stimson on April 25, concerned S-l (the Manhattan Project). Stimson, with Groves present during part of the meeting, traced the history of the Manhattan Project, summarized its status, and detailed the timetable for testing and combat delivery. Truman asked numerous questions during the forty-five-minute meeting and made it clear that he understood the relevance of the atomic bomb to upcoming diplomatic and military initiatives.

By the time Truman took office, Japan was near defeat. American aircraft were attacking Japanese cities at will. A single firebomb raid in March killed nearly 100,000 people and injured over a million in Tokyo. A second air attack on Tokyo in May killed 83,000. Meanwhile, the United States Navy had cut the islands' supply lines. But because of the generally accepted view that the Japanese would fight to the bitter end, a costly invasion of the home islands seemed likely, though some American policy makers held that successful combat delivery of one or more atomic bombs might convince the Japanese that further resistance was futile.

The Interim Committee Report

On June 6, Stimson again briefed Truman on S-l. The briefing summarized the consensus of an Interim Committee meeting held on May 31. The Interim Committee was an advisory group on atomic research composed of Bush, Conant, Karl Compton, Under Secretary of the Navy Ralph A. Bard, Assistant Secretary of State William L. Clayton, and future Secretary of State James F. Byrnes. Oppenheimer, Fermi, Arthur Compton, and Lawrence served as scientific advisors (the Scientific Panel), while Marshall represented the military. A second meeting on June 1 with Walter S. Carpenter of DuPont, James C. White of Tennessee Eastman, George H. Bucher of Westinghouse, and James A. Rafferty of Union Carbide provided input from the business side. The Interim Committee was charged with recommending the proper use of atomic weapons in wartime and developing a position for the United States on postwar atomic policy. The May 31 meeting concluded that the United States should try to retain superiority of nuclear weapons in case international relations deteriorated.[45] Most present at the meeting thought that the United States should protect its monopoly for the present, though they conceded that the secrets could not be held long. It was only a matter of time before another country, presumably Russia, would be capable of producing atomic weapons.

There was some discussion of free exchange of nuclear research for peaceful purposes and the

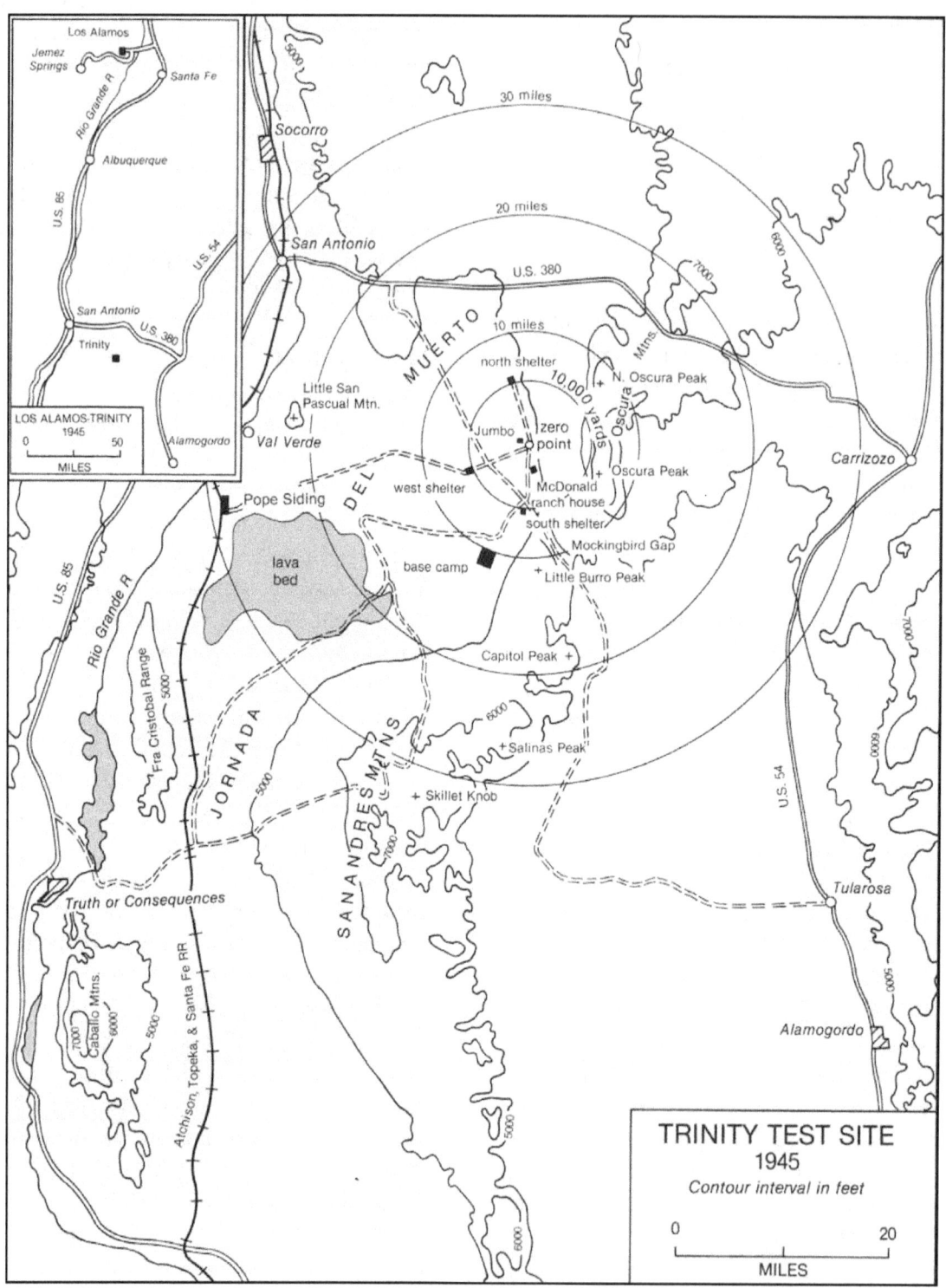

Trinity Test Site. Reprinted from Vincent C. Jones, *Manhattan: The Army and the Atomic Bomb* (Washington, D.C.: U.S. Government Printing Office, 1985).

international inspection system that such an exchange would require. Lawrence's suggestion that a demonstration of the atomic bomb might possibly convince the Japanese to surrender was discussed over lunch and rejected. The bomb might be a dud, the Japanese might put American prisoners of war in the area, or shoot down the plane, and the shock value of the new weapon would be lost. These reasons and others convinced the group that the bomb should be dropped without warning on a dual target—a war plant surrounded by workers' homes. The meeting with the industrialists on June 1 further convinced the Interim Committee that the United States had a lead of three to ten years on the Soviet Union in production facilities for bomb fabrication.

On June 6, Stimson informed the President that the Interim Committee recommended keeping S-1 a secret until Japan had been bombed. The attack should take place as soon as possible and without warning. Truman and Stimson agreed that the President would stall if approached about atomic weapons in Berlin, but that it might be possible to gain concessions from Russia later in return for providing technical information. Stimson told Truman that the Interim Committee was considering domestic legislation and that its members generally held the position that international agreements should be made in which all nuclear research would be made public and a system of inspections would be devised. In case international agreements were not forthcoming, the United States should continue to produce as much fissionable material as possible to take advantage of its current position of superiority.

Planning for Surrender

Strategies for forcing Japanese capitulation occupied center stage in June. Truman gained Chinese concurrence in the Yalta agreements by assuring T. V. Soong, the Chinese foreign minister, that Russia's intentions in the Far East were benevolent, smoothing the way for the entrance

of the Red Army. Joseph C. Grew, acting secretary of state, clarified the definition of unconditional surrender. Japan need not fear total annihilation, Grew stated. Once demilitarized, Japan would be free to choose its political system and would be allowed to develop a vibrant economy. Grew hoped that a public statement to Japan would lead to surrender before a costly invasion would have to be launched. The Joint Chiefs of Staff continued to advocate the invasion of Kyushu, a plan identified as Operation Olympic. Stimson hoped that an invasion could be avoided, either by redefining the surrender terms or by using the atomic bomb.

Indicative of the wide range of his responsibilities was Groves's position as head of a bomb target selection group set up in late April, a responsibility he shared with General Thomas Farrell, appointed Groves's military aide in February 1945. In late May, the committee of scientists and Army Air Force officers listed Kokura Arsenal, Hiroshima, Niigata, and Kyoto as the four best targets, believing that attacks on these cities—none of which had yet been bombed by Curtis LeMay's Twentieth Air Force (which planned to eliminate all major Japanese cities by January 1, 1946)—would make a profound psychological impression on the Japanese and weaken military resistance. Stimson vetoed Kyoto, Japan's most cherished cultural center, and Nagasaki replaced the ancient capital in the directive issued to the Army Air Force on July 25.[46]

The Franck Report and Its Critics

Meanwhile the Met Lab was beginning to stir. The Scientific Panel of the Interim Committee was the connection between the scientists and the policy makers, and Compton was convinced that there must be a high level of participation in the decision-making process. His June 2 briefing of the Met Lab staff regarding the findings of the Interim Committee led to a flurry of activity. The Met Lab's Committee on the Social and Political

Tower For Trinity Test. *Department of Energy.*

Trinity Device Being Readied. Reprinted from Richard G. Hewlett and Oscar E. Anderson, Jr., *The New World, 1939-1946,* Volume I of *A History of the United States Atomic Energy Commission* (University Park: Pennsylvania State University Press, 1962).

Implications of the Atomic Bomb, chaired by James Franck and including Seaborg and Szilard, issued a report advocating international control of atomic power as the only way to stop the arms race that would be inevitable if the United States bombed Japan without first demonstrating the weapon in an uninhabited area.

The Scientific Panel disagreed with the Franck Report, as the Met Lab study was known, and concluded that no technical test would convince Japan to surrender. The Panel concluded that such a military demonstration of the bomb might best further the cause of peace but held that such a demonstration should take place only after the United States informed its allies. On June 21, the Interim Committee sided with the

position advanced by the Scientific Panel. The bomb should be used as soon as possible, without warning, and against a war plant surrounded by additional buildings. As to informing allies, the Committee concluded that Truman should mention that the United States was preparing to use a new kind of weapon against Japan when he went to Berlin in July. On July 2, 1945, President Truman listened as Stimson outlined the peace terms for Japan, including demilitarization and prosecution of war criminals in exchange for economic and governmental freedom of choice. Stimson returned on July 3 and suggested that Truman broach the issue of the bomb with Stalin and tell the Soviet leader that it could become a force for peace with proper agreements.[47]

Remains of Trinity Test Tower Footings. Oppenheimer and Groves at Center. *Department of Energy.*

The Trinity Test

Meanwhile, the test of the plutonium weapon, named Trinity by Oppenheimer (a name inspired by the poems of John Donne), was rescheduled for July 16 at a barren site on the Alamogordo Bombing Range known as the Jornada del Muerto, or Journey of Death, 210 miles south of Los Alamos. A test explosion had been conducted on May 7 with a small amount of fissionable material to check procedures and fine-tune equipment. Preparations continued through May and June and were complete by the beginning of July. Three observation bunkers located 10,000 yards north, west, and south of the firing tower at ground zero would attempt to measure critical aspects of the reaction. Specifically, scientists would try to determine the symmetry of the implosion and the amount of energy released. Additional measurements would be taken to determine damage estimates, and equipment would record the behavior of the fireball. The biggest concern was control of the radioactivity the test device would release. Not entirely content to trust favorable meteorological conditions to carry the radioactivity into the upper atmosphere, the Army stood ready to evacuate the people in surrounding areas.

On July 12, the plutonium core was taken to the test area in an army sedan. The non-nuclear components left for the test site at 12:01 a.m., Friday the 13th. During the day on the 13th, final assembly of the gadget took place in the McDonald ranch house. By 5:00 p.m. on the 15th, the device had been assembled and hoisted atop the one-hundred-foot firing tower. Groves, Bush, Conant, Lawrence, Farrell, Chadwick (head of the British contingent at Los Alamos and discoverer of the neutron), and others arrived in the test area, where it was pouring rain. Groves and Oppenheimer, standing at the S-10,000 control bunker, discussed what to do if the weather did not break in time for the scheduled 4:00 a.m. test. At 3:30 they pushed the time back to 5:30; at 4:00 the rain stopped. Kistiakowsky and his team armed the device shortly after 5:00 a.m. and retreated to S-10,000. In accordance with his policy that each observe from different locations in case of an accident, Groves left Oppenheimer and joined Bush and Conant at base camp. Those in shelters heard the countdown over the public address system, while observers at base camp picked it up on an FM radio signal.[48]

The Dawn of the Atomic Age

At precisely 5:30 a.m. on Monday, July 16, 1945, the atomic age began. While Manhattan staff members watched anxiously, the device exploded over the New Mexico desert, vaporizing the tower and turning asphalt around the base of the tower to green sand. The bomb released approximately 18.6 kilotons of power, and the New Mexico sky was suddenly brighter than many suns. Some observers suffered temporary blindness even though they looked at the brilliant light through smoked glass. Seconds after the explosion came a huge blast, sending searing heat across the desert and knocking some observers to the ground. A steel container weighing over 200 tons, standing a half-mile from ground zero, was knocked ajar. (Nicknamed Jumbo, the huge container had been ordered for the plutonium test and transported to the test site but eliminated during final planning.)

As the orange and yellow fireball stretched up and spread, a second column, narrower than the first, rose and flattened into a mushroom shape, thus providing the atomic age with a visual image that has become imprinted on the human consciousness as a symbol of power and awesome destruction.[49]

At base camp, Bush, Conant, and Groves shook hands. Oppenheimer reported later that the experience called to his mind the legend of Prometheus, punished by Zeus for giving man fire. He also thought fleetingly of Alfred Nobel's vain hope that dynamite would end wars. The terrifying destructive power of atomic weapons and the uses to which they might be put were to haunt many of the Manhattan Project scientists for the remainder of their lives.[50]

The success of the Trinity test meant that a second type of atomic bomb could be readied for use against Japan. In addition to the uranium gun model, which was not tested prior to being used in combat, the plutonium implosion device detonated at Trinity now figured in American Far Eastern strategy. In the end Little Boy, the untested uranium bomb, was dropped first at Hiroshima on August 6, 1945, while the plutonium weapon Fat Man followed three days later at Nagasaki on August 9.

Potsdam
The American contingent to the Big Three conference, headed by Truman, Byrnes, and Stimson, arrived in Berlin on July 15 and spent most of the next two days grappling with the interrelated issues of Russian participation in the Far Eastern conflict and the wording of an early surrender offer that might be presented to the Japanese. This draft surrender document received considerable attention, the sticking point being the term "unconditional." It was clear that the Japanese would fight on rather than accept terms that would eliminate the Imperial House or demean the warrior tradition, but American

policy makers feared that anything less than a more democratic political system and total demilitarization might lead to Japanese aggression in the future. Much effort went into finding the precise formula that would satisfy American war aims in the Pacific without requiring a costly invasion of the Japanese mainland. In an attempt to achieve surrender with honor, the emperor had instructed his ministers to open negotiations with Russia. The United States intercepted and decoded messages between Tokyo and Moscow that made it unmistakably clear that the Japanese were searching for an alternative to unconditional surrender.

Reports on Trinity
Stalin arrived in Berlin a day late, leaving Stimson July 16 to mull over questions of postwar German administration and the Far Eastern situation. After sending Truman and Byrnes a memorandum advocating an early warning to Japan and setting out a bargaining strategy for Russian entry in the Pacific war, Stimson received a cable from George L. Harrison, his special consultant in Washington, that read:

> Operated on this morning. Diagnosis not yet complete but results seem satisfactory and already exceed expectations. Local press release necessary as interest extends great distance. Dr. Groves pleased. He returns tomorrow. I will keep you posted.[51]

Stimson immediately informed Truman and Byrnes that the Trinity test had been successful. The next day Stimson informed Churchill of the test. The prime minister expressed great delight and argued forcefully against informing the Russians, though he later relented. On July 18, while debate continued over the wording of the surrender message, focusing on whether or not to guarantee the place of the emperor, Stimson received a second cable from Harrison:

Doctor has just returned most enthusiastic and confident that the little boy is as husky as his big brother. The light in his eyes discernible from here to Highhold and I could have heard his screams from here to my farm.[52]

Translation: Groves thought the plutonium weapon would be as powerful as the uranium device and that the Trinity test could be seen as far away as 250 miles and the noise heard for fifty miles. Initial measurements taken at the Alamogordo site suggested a yield in excess of 5,000 tons of TNT. Truman went back to the bargaining table with a new card in his hand.

Further information on the Trinity test arrived on July 21 in the form of a long and uncharacteristically excited report from Groves's. Los Alamos scientists now agreed that the blast had been the equivalent of between 15,000 and 20,000 tons of TNT, higher than anyone had predicted. Groves reported that glass shattered 125 miles away, that the fireball was brighter than several suns at midday, and that the steel tower had been vaporized. Though he had previously believed it impregnable, Groves stated that he did not consider the Pentagon safe from atomic attack.[53] Stimson informed Marshall and then read the entire report to Truman and Byrnes. Stimson recorded that Truman was "tremendously pepped up" and that the document gave him an entirely new feeling of confidence.[54] The next day Stimson, informed that the uranium bomb would be ready in early August, discussed Grove's report at great length with Churchill. The British prime minister was elated and said that he now understood why Truman had been so forceful with Stalin the previous day, especially in his opposition to Russian designs on Eastern Europe and Germany. Churchill then told Truman that the bomb could lead to Japanese surrender without an invasion and eliminate the necessity for Russian military help. He recommended that the President continue to take a hard line with

Stalin. Truman and his advisors shared Churchill's views. The success of the Trinity test stiffened Truman's resolve, and he refused to accede to Stalin's new demands for concessions in Turkey and the Mediterranean.

On July 24, Stimson met with Truman. He told the President that Marshall no longer saw any need for Russian help, and he briefed the President on the latest S-l situation. The uranium bomb might be ready as early as August 1 and was a certainty by August 10. The plutonium weapon would be available by August 6. Stimson continued to favor making some sort of commitment to the Japanese emperor, though the draft already shown to the Chinese was silent on this issue.

Truman Informs Stalin

American and British coordination for an invasion of Japan continued, with November 1 standing as the landing date. At a meeting with American and British military strategists at Potsdam, the Russians reported that their troops were moving into the Far East and could enter the war in mid-August. They would drive the Japanese out of Manchuria and withdraw at the end of hostilities. Nothing was said about the bomb. This was left for Truman, who, on the evening of July 24, approached Stalin without an interpreter to inform the Generalissimo that the United States had a new and powerful weapon. Stalin casually responded that he hoped that it would be used against Japan to good effect. The reason for Stalin's composure became clear later when it was learned that Russian intelligence had been receiving information about the S-l project from Klaus Fuchs and other agents since summer 1942.

The Potsdam Proclamation

A directive, written by Groves and issued by Stimson and Marshall on July 25, ordered the Army Air Force's 509th Composite Group to attack Hiroshima, Kokura, Niigata, or Nagasaki

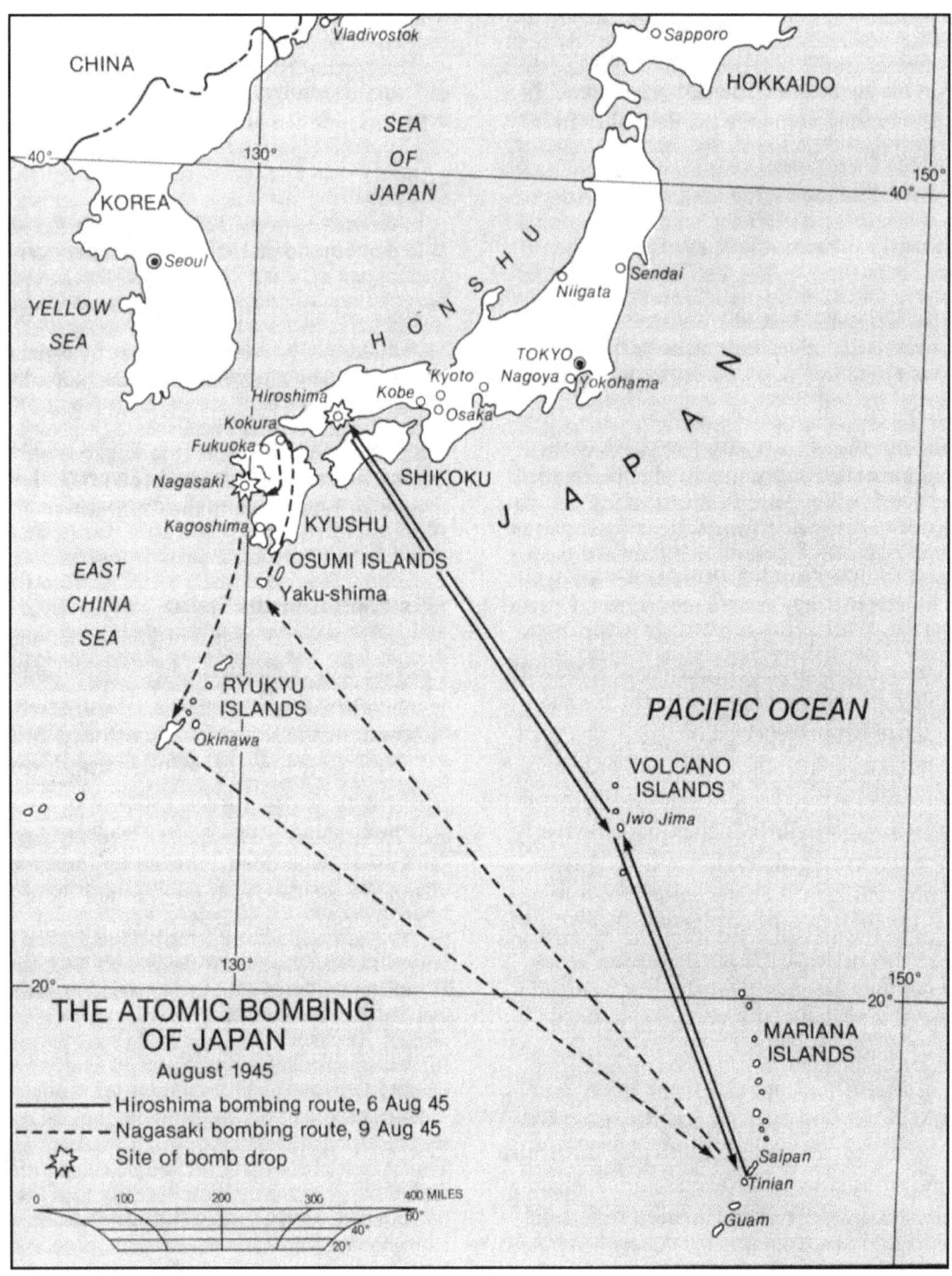

Atomic Bombing of Japan. Reprinted from Vincent C. Jones, *Manhattan: The Army and the Atomic Bomb* (Washington, D.C.: U.S. Government Printing Office, 1985).

Model of Little Boy Uranium Bomb. Reprinted from Richard G. Hewlett and Oscar E. Anderson, Jr., *The New World 1939-1946* Volume I of *A History of the United States Atomic Energy Commission* (University Park: Pennsylvania State University Press, 1962).

"after about" August 3, or as soon as weather permitted.[55] The 509th was ready. Tests with dummies had been conducted successfully, and Operation Bronx, which brought the gun and uranium-235 projectile to Tinian aboard the U.S.S. Indianapolis and the other components on three C-54s, was complete. On July 26, the United States learned of Churchill's electoral defeat and Chiang Kai-Shek's concurrence in the warning to Japan. Within hours the warning was issued in the name of the President of the United States, the president of China, and the prime minister of Great Britain (now Clement Attlee). The Russians were not informed in advance. This procedure was technically correct since the Russians were not at war with Japan, but it was another indication of the new American attitude that the Soviet Union's aid in the present conflict no longer was needed. The message called for the Japanese to surrender unconditionally or face "prompt and utter destruction."[56] The Potsdam Proclamation left the emperor's status unclear by making no reference to the royal house in the section that

promised the Japanese that they could design their new government as long as it was peaceful and more democratic. While anti-war sentiment was growing in Japanese decision-making circles, it could not carry the day as long as unconditional surrender left the emperor's position in jeopardy. The Japanese rejected the offer on July 29.

Intercepted messages between Tokyo and Moscow revealed that the Japanese wanted to surrender but felt they could not accept the terms offered in the Potsdam Proclamation. American policy makers, however, anxious to end the war without committing American servicemen to an invasion of the Japanese homeland, were not inclined to undertake revisions of the unconditional surrender formula and cause further delay. A Russian declaration of war might convince Japan to surrender, but it carried a potentially prohibitive price tag as Stalin would expect to share in the postwar administration of Japan, a situation that would threaten American plans in the Far East. A blockade of Japan combined with conventional bombing was rejected as too time-consuming and an invasion of the islands as too costly. And few believed that a demonstration of the atomic bomb would convince the Japanese to give up. Primarily upon these grounds, American policy makers concluded that the atomic bomb must be used. Information that Hiroshima might be the only prime target city without American prisoners in the vicinity placed it first on the list. As the final touches were put on the message Truman would issue after the attack, word came that the first bomb could be dropped as early as August 1. With the end now in sight, poor weather led to several days' delay.

Hiroshima

In the early morning hours of August 6, 1945, a B-29 bomber attached to the 509th Composite Group took off from Tinian Island and headed north by northwest toward the Japanese Islands over 1,500 miles away. Its primary target was Hiroshima, an important military and communications center with a population of nearly 300,000 located in the deltas of southwestern Honshu Island facing the Inland Sea. The *Enola Gay*, piloted by Colonel Paul Tibbets, flew at low altitude on automatic pilot

Fat Man Plutonium Bomb Being Readied at Tinian.
Los Alamos National Laboratory.

before climbing to 31,000 feet as it neared the target area. As the observation and photography escorts dropped back, the *Enola Gay* released a 9,700-pound uranium bomb, nicknamed Little Boy, at approximately 8:15 a.m. Hiroshima time. Tibbets immediately dove away to avoid the anticipated shockwaves of the blast. Forty-three seconds later a huge explosion lit the morning sky as Little Boy detonated 1900 feet above the city, directly over a parade field where the Japanese Second Army was doing calisthenics. Though already eleven and a half miles away, the Enola Gay was rocked by the blast. At first Tibbets thought he was taking flak. After a second shockwave hit the plane, the crew looked back at Hiroshima. "The city was hidden by that awful cloud boiling up, mushrooming, terrible and incredibly tall," Tibbets recalled.[57] Little Boy killed 70,000 people (including about twenty American airmen being held as POWs) and injured another 70,000. By the end of 1945, the Hiroshima death toll rose to 140,000 as radiation-sickness deaths mounted. Five years later the total reached 200,000. The bomb caused total devastation for five square miles, with almost all of the buildings in the city either destroyed or damaged.

Within hours of the attack, radio stations began reading a prepared statement from President Harry Truman informing the American public that the United States had dropped an entirely new type of bomb on the Japanese city of Hiroshima—an atomic bomb with more power than 15,000 tons of TNT.[58] Truman warned that if Japan still refused to surrender unconditionally as demanded by the Potsdam Proclamation of July 26, the United States would attack additional targets with equally devastating results. Two days later, on August 8, the Soviet Union declared war on Japan and attacked Japanese forces in Manchuria, ending American hopes that the war would end before Russian entry into the Pacific theater.

Nagasaki

Factional struggles and communications problems prevented Japan from meeting Allied terms in the immediate aftermath of Hiroshima. In the absence of a surrender announcement, conventional bombing raids on additional Japanese cities continued as scheduled. Then, on August 9, a second atomic attack took place. Taking off from Tinian at 3:47 a.m., *Bock's Car* (named after its usual pilot) headed for its primary target, Kokura Arsenal, located on the northern coast of Kyushu Island. Pilot Charles Sweeney found unacceptable weather conditions and unwelcome flak above Kokura. Sweeney made three passes over Kokura, then decided to switch to his secondary target even though he had only enough fuel remaining for a single bombing run. Clouds greeted *Bock's Car* as it approached Nagasaki, home to the Mitsubishi plant that had manufactured the torpedoes used at Pearl Harbor. At the last minute, a brief break in the cloud cover made possible a visual targeting at 29,000 feet and *Bock's Car* dropped her single payload, a plutonium bomb weighing 10,000 pounds and nicknamed Fat Man, at 11:01 a.m. The plane then veered off and headed to Okinawa for an emergency landing. Fat Man exploded 1,650 feet above the slopes of the city with a force of 21,000 tons of TNT.[59] Fat Man killed 40,000 people and injured 60,000 more. Three square miles of the city were destroyed, less than Hiroshima because of the steep hills surrounding Nagasaki. By January 1946, 70,000 people had died in Nagasaki. The total eventually reached 140,000, with a death rate similar to that of Hiroshima.[60]

Surrender

Still the Japanese leadership struggled to come to a decision, with military extremists continuing to advocate a policy of resistance to the end. Word finally reached Washington from Switzerland and Sweden early on August 10 that the Japanese, in accordance with Hirohito's wishes, would accept the surrender terms, provided the emperor retain his position. Truman held up a third

atomic attack while the United States considered a response, finally taking a middle course and acknowledging the emperor by stating that his authority after the surrender would be exercised under the authority of the Supreme Commander of the Allied Powers. With British, Chinese, and Russian concurrence, the United States answered the Japanese on August 11. Japan surrendered on August 14, 1945, ending the war that began for the United States with the surprise attack at Pearl Harbor on December 7, 1941. The United States had been celebrating for almost three weeks when the formal papers were signed aboard the U.S.S. Missouri on September 2.

The Bomb Goes Public

The veil of secrecy that had hidden the atomic bomb project was lifted on August 6 when President Truman announced the Hiroshima raid to the American people. The release of the Smyth Report on August 12, which contained general technical information calculated to satisfy public curiosity without disclosing any atomic secrets, brought the Manhattan Project into fuller view.[61] Americans were astounded to learn of the existence of a far-flung, government-run, top secret operation with a physical plant, payroll, and labor force comparable in size to the American automobile industry. Approximately 130,000 people were employed by the project at its peak, among them many of the nation's leading scientists and engineers.

In retrospect, it is remarkable that the atomic bomb was built in time to be used in World War II. Most of the theoretical breakthroughs in nuclear physics dated back less than twenty-five years, and with new findings occurring faster than they could be absorbed by practitioners in the field, many fundamental concepts in nuclear physics and chemistry had yet to be confirmed by laboratory experimentation. Nor was there any conception initially of the design and engineering difficulties that would be involved in translating what was known theoretically into working devices capable of releasing the enormous energy of the atomic nucleus in a predictable fashion. In fact, the Manhattan Project was as much a triumph of engineering as of science. Without the innovative work of the talented Leslie Groves, as well as that of Crawford Greenewalt of DuPont and others, the revolutionary breakthroughs in nuclear science achieved by Enrico Fermi, Niels Bohr, Ernest Lawrence, and their colleagues would not have produced the atomic bomb during World War II. Despite numerous obstacles, the United States was able to combine the forces of science, government, military, and industry into an organization that took nuclear physics from the laboratory and into battle with a weapon of awesome destructive capability, making clear the importance of basic scientific research to national defense.

Part VI:
The Manhattan District in Peacetime

From the time S-l became public knowledge until the Atomic Energy Commission succeeded it on January 1, 1947, the Manhattan Engineer District controlled the nation's nuclear program. Groves remained in command, intent upon protecting America's lead in nuclear weapons by completing and consolidating the organization he had presided over for three years in challenging wartime conditions. He soon found that peacetime held its own challenges.

According to a plan approved by Stimson and Marshall in late August 1945, Groves shut down the thermal diffusion plant in the K-25 area on September 9 and put the Alpha tracks at Y-12 on standby during September as well. The improved K-25 gaseous diffusion plant now provided feed directly to the Beta units. Hanford's three piles continued in operation, but one of the two chemical separation areas was closed. Los Alamos was assigned the task of producing a stockpile of atomic weapons. Actual weapon assembly was to be done at Sandia Base in Albuquerque, where engineering and technical personnel were relocated with the staff previously stationed at Wendover Field in western Utah.

Operation Crossroads
In July 1946, during Operation Crossroads, the Manhattan Project tested its third and fourth plutonium bombs (Trinity and Nagasaki were the first two) with a large, invited audience of journalists, scientists, military officers, congressmen, and foreign observers at Bikini

Atoll in the Pacific. Shot Able, dropped from a B-29 on July 1, sank three ships and performed as well as its two predecessors from a technical standpoint, though it failed to fulfill its pretest publicity buildup. Shot Baker was detonated from ninety feet underwater on the morning of July 25. Baker produced a spectacular display as it wreaked havoc on a seventy-four-vessel fleet of empty ships and spewed thousands of tons of water into the air. Both Able and Baker yielded explosions equivalent to 21,000 tons of TNT, though Baker introduced the most subtle hazard of the atomic age—radiation fallout.[62] Able and Baker were the final weapon tests conducted by the Manhattan Project and the last American tests until the Atomic Energy Commission's Sandstone series began in spring 1948.

Superpower Chill
Between August 1945 and January 1947, while Groves fought to maintain the high priority of the atomic program in a peacetime environment, the euphoria that swept the United States at the end of World War II dissipated as Americans found themselves embroiled in a new global struggle, this time with the Soviet Union. The United States held a monopoly on atomic weapons during the sixteen months of Groves's peacetime tenure, but less than three years after the Atomic Energy Commission succeeded the Manhattan Engineer District, the Russians' secret atomic bomb program achieved success with the 1949 test of Joe I (which the Americans named after Joseph Stalin). During the 1950s, relations between the two superpowers remained strained, and both added the hydrogen bomb to their arsenals in an attempt to achieve military superiority.

Postwar Planning
The beginning of the Cold War in the late 1940s was linked to the failure of the World War II allies to reach agreements on international controls respecting nuclear research and atomic weapons. Postwar planning in the United States began in earnest in July 1944, when Met Lab scientists in

Chicago issued a "Prospectus on Nucleonics," which included plans for atomic research and advocated the creation of an international organization to prevent nuclear conflict. In August the Military Policy Committee set up a Postwar Policy Committee, charged with making recommendations on the proper government role in postwar atomic research and development. The committee, composed of Richard Tolman (chairman), Warren Lewis, Henry Smyth, and Rear Admiral Earle W. Mills, recommended that the best way for the government to maintain a vigorous nuclear program was to set up a peacetime version of the Office of Scientific Research and Development.

Roosevelt and Churchill included postwar planning on their agenda when they met at Hyde Park in September 1944. They immediately vetoed the idea of an open atomic world (Churchill adamantly rejecting Bohr's recommendation.) Bush and Conant, meanwhile, contacted Stimson on September 19 and spoke to the necessity of releasing selected information on the bomb project, reasoning that in a free country the secret could not be kept long. When Roosevelt asked Bush for a briefing on S-l several days later, Bush discovered that Roosevelt had signed an "aide-memoire" with Churchill, pledging to continue bilateral research with England in certain areas of atomic technology.[63] Bush feared that Roosevelt would institute full interchange with Great Britain without consulting his own atomic power experts. Bush argued, prophetically, that leaving the Russians out of such an arrangement might well lead to an arms race among the Allied victors.

The Baruch Plan

Bush and Conant presented their views more fully on September 30. They held that the American and British lead would last no more than three or four years and that security against the bigger bombs that surely would result from a worldwide arms race could be gained only through

international agreements aimed at preventing secret research and surprise attacks. Bush and Conant's basic philosophy found expression in the Acheson-Lilienthal report of March 1946, fashioned primarily by Oppenheimer and evolving into the formal American proposal for the international control of atomic energy known as the Baruch Plan.

Bernard Baruch, the elder statesman who had served American presidents in various capacities since World War I, unveiled the United States plan in a speech to the newly-created United Nations Atomic Energy Commission on June 14, 1946. Baruch proposed the establishment of an international atomic development authority along the lines proposed by the Acheson-Lilienthal report, one that would control all activities dangerous to world security and possess the power to license and inspect all other nuclear projects. Once such an authority was established, no more bombs should be built and existing bombs should be destroyed. Abolishing atomic weapons could lay the groundwork for reducing and subsequently eliminating all weapons, thus outlawing war altogether. The Baruch Plan, in Baruch's words "the last, best hope of earth," deviated from the optimistic tone of the Acheson-Lilienthal plan, which had intentionally remained silent on enforcement, and set specific penalties for violations such as illegally owning atomic bombs.[64] Baruch argued that the United Nations should not allow members to use the veto to protect themselves from penalties for atomic energy violations; he held that simple majority rule should prevail in this area. As on enforcement, the Acheson-Lilienthal report had studiously avoided comment on the veto issue.[65]

Not surprisingly, the Soviet Union, a non-nuclear power, insisted upon retaining its United Nations veto and argued that the abolition of atomic weapons should precede the establishment of an international authority. Negotiations could not proceed fairly, the Russians maintained, as long as the United States could use its atomic monopoly to coerce other nations into accepting its plan. The Baruch Plan proposed that the United States reduce its atomic arsenal by carefully defined stages linked to the degree of international agreement on control. Only after each stage of international control was implemented would the United States take the next step in reducing its stockpile. The United States position, then, was that international agreement must precede any American reductions, while the Soviets maintained that the bomb must be banned before meaningful negotiations could take place.

The debate in the United Nations was a debate in name only; neither side budged an inch in the six months following Baruch's United Nations speech. In the end, the Soviet Union, unwilling to surrender its veto power, abstained from the December 31, 1946, vote on Baruch's proposal on the grounds that it did not prohibit the bomb, and the American plan became a dead letter by early 1947—though token debate on the American plan continued into 1948. The United States, believing that Soviet troops posed a threat to eastern Europe and recognizing that American conventional forces were rapidly demobilizing, refused to surrender its atomic deterrent without adequate international controls and continued to develop its nuclear arsenal. In an atmosphere of mutual suspicion the Cold War set in.

The Debate Over Domestic Legislation

While the international situation grew more ominous due to deteriorating relations between the United States and the Soviet Union, a domestic debate was taking place over the permanent management of America's nuclear program. The terms of the debate were framed by the Interim Committee in July 1945 when it wrote draft legislation proposing a peacetime organization with responsibilities very similar to those of the Manhattan Project. The draft legislation provided for a strong military presence on a nine-member board of commissioners and strongly advocated the federal government's continued dominance in nuclear research and development.

The May-Johnson Bill

The Interim Committee's draft legislation reached President Truman via the State Department shortly after the armistice. After affected federal agencies approved, Truman advocated speedy passage of the congressional version of the bill, the May-Johnson bill, on October 3, 1945. Groves, Bush, and Conant testified at hearings in the House of Representatives that the sweeping powers granted the proposed commission were necessary and that only government control of atomic power could prevent its misuse. Although Lawrence, Fermi, and Oppenheimer (with some misgivings) regarded the bill as acceptable, many of the scientists at the Met Lab and at Oak Ridge complained that the bill was objectionable because it was designed to maintain military control over nuclear research, a situation that had been tolerable during the war but was unacceptable during peacetime when free scientific interchange should be resumed. Particularly onerous to the scientific opponents were the proposed penalties for security violations contained in the May-Johnson bill—ten years in prison and a $100,000 fine. Organized scientific opposition in Washington slowed the bill's progress, and Arthur H. Vandenberg of Michigan held it up in the Senate through a parliamentary maneuver.

The McMahon Bill

As support for the May-Johnson bill eroded in late 1945, President Truman withdrew his support. Vandenberg's attempt to establish a joint House-Senate special committee failed, but Brien McMahon of Connecticut successfully created and became chair of the Senate's Special Committee on Atomic Energy. Daily hearings took place until December 20, when McMahon introduced a substitute to the May-Johnson bill. Hearings on the new McMahon bill began in January. Groves opposed McMahon's bill, citing weak security provisions, the low military presence, and the stipulation that commission members be full-time (Groves thought that more eminent commissioners could be obtained if work was part-time). Groves also objected to the bill's provision that atomic weapons be held in civilian rather than military custody. Nevertheless, the Senate approved the McMahon bill on June 1, 1946, and the House approved it on July 20, with a subsequent conference committee eliminating most substantive amendments. The sometimes bitter debate between those who advocated continued military stewardship of America's atomic arsenal and those who saw continued military control as inimical to American traditions ended in victory for supporters of civilian authority. President Truman signed the McMahon Act, known officially as the Atomic Energy Act of 1946, on August 1. The bill called for the transfer of authority from the United States Army to the United States Atomic Energy Commission, a five-member

civilian board serving full-time and assisted by a general advisory committee and a military liaison committee.[66] The Atomic Energy Act entrusted the Atomic Energy Commission with the government monopoly in the field of atomic research and development previously held by its wartime predecessor.[67]

Conclusion

The Manhattan Project, its wartime mission completed, gave way to the civilian Atomic Energy Commission. How well the Atomic Energy Commission would be able to manage the nuclear arsenal in a Cold War environment and whether it could successfully develop the peaceful uses of atomic energy, only time would tell. What was clear as the Atomic Energy Commission took over at the beginning of 1947 was that the success of the Manhattan Project had helped cement the bond between basic scientific research and national security. Science had gone to war and contributed mightily to the outcome. The challenge confronting American policy makers in the postwar years was to enlist the forces of science in the battle to defend the peace.

Manhattan Project Chart

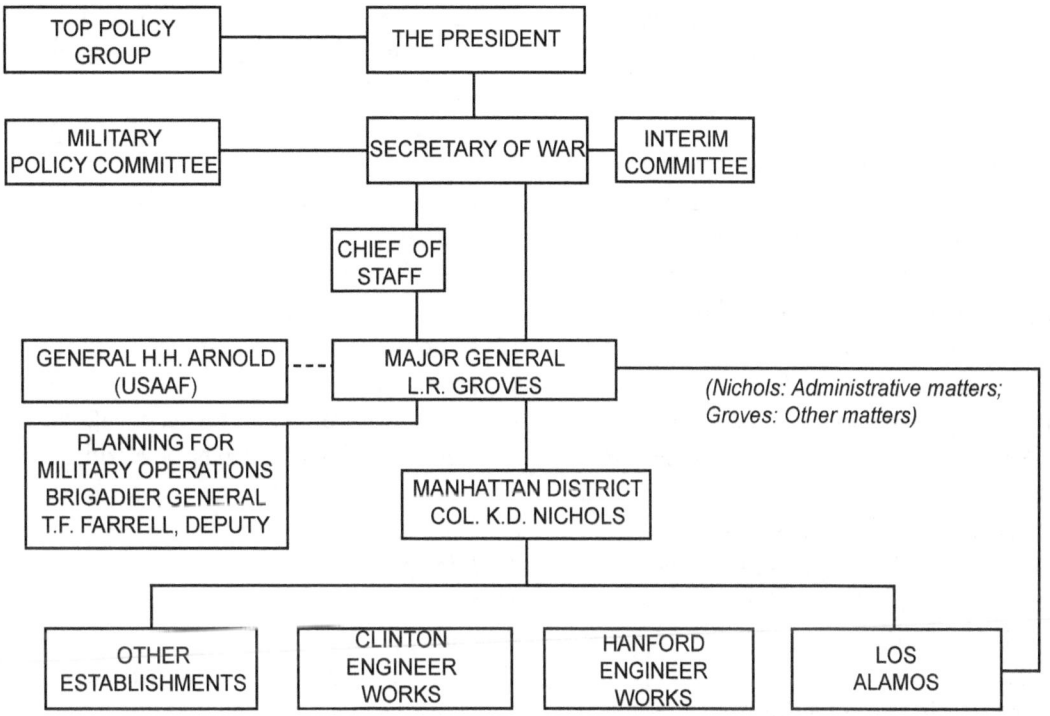

Simplified Chart of Manhattan Project. Adapted from Leslie R. Groves, *Now It Can Be Told* (New York: Harper & Row, 1962).

Notes

1. The Einstein letter is reprinted in Vincent C. Jones, *Manhattan: The Army and the Atomic Bomb* (Washington, D.C.: U.S. Government Printing Office, 1985), pp. 609-10.
2. Roosevelt to Einstein, October 19, 1939.
3. The United States had little reliable intelligence on the German bomb effort until late in the war. Thus it was not known until the ALSOS counterintelligence mission that the German program had not proceeded beyond the laboratory stage and had foundered by mid-1942. For details of the ALSOS mission see Richard Rhodes, *The Making of the Atomic Bomb* (New York: Simon & Schuster, 1986), pp. 605-10.
4. For details on the German research effort see McGeorge Bundy, *Danger and Survival: Choices About the Bomb in the First Fifty Years* (New York: Random House, 1988), pp. 14-23.
5. Lawrence Badash, "Introduction," in *Reminiscences of Los Alamos, 1943-1945,* edited by Lawrence Badash, Joseph O. Hirschfelder, and Herbert P. Broida (Dordrecht, Holland: D. Reidel Publishing Company, 1980), xi.
6. Rhodes, *Making of the Atomic Bomb*, p. 228.
7. William R. Shea, "Introduction: From Rutherford to Hahn," in *Otto Hahn and the Rise of Nuclear Physics*, edited by William R. Shea (Dordrecht, Holland: D. Reidel Publishing Company, 1983), p. 15.
8. *Ibid.*
9. Jones, *Manhattan*, p. 10.
10. Richard G. Hewlett and Oscar E. Anderson, Jr., *The New World, 1939-1946, Volume I, A History of the United States Atomic Energy Commission* (University Park, Pennsylvania: Pennsylvania State University Press, 1962), pp. 30-31.
11. *Ibid.*, p. 168.
12. Laura Fermi, *Atoms in the Family: My Life With Enrico Fermi* (Chicago: University of Chicago Press, 1954), p. 164.
13. Rhodes, *Making of the Atomic Bomb*, pp. 336-38.
14. Hewlett and Anderson, *New World*, p. 37.
15. *Ibid.*, p. 39.
16. MAUD, while it appears to be an acronym, is not. It is simply a codename.
17. Rhodes, *Making of the Atomic Bomb*, pp. 321-25 and 369.
18. Bundy, *Danger and Survival*, pp. 48-49.
19. Hewlett and Anderson, New World, p. 46.
20. *Ibid.*, p. 49.
21. *Ibid.*, p. 48.
22. *Ibid.*, p. 406.
23. *Ibid.*, pp. 74-75.
24. *Ibid.*, pp. 82-83.
25. Henry D. Smyth, *A General Account of the Development of Methods of Using Atomic Energy for Military Purposes Under the Auspices of the United States Government, 1940-1945* (Washington, D.C.: U. S. Government Printing Office, 1945), p. 70.
26. Rhodes, *Making of the Atomic Bomb*, p. 442.
27. *Ibid.*
28. Hewlett and Anderson, *New World*, p. 113.
29. Kenneth D. Nichols, *The Road to Trinity* (New York: William Morrow and Company, Inc., 1987), pp. 121 and 146.
30. Hewlett and Anderson, *New World*, p. 149.
31. Nichols recounts his adventure in borrowing the silver in *Road to Trinity*, p. 42.
32. Hewlett and Anderson, *New World*, p. 155.

33. *Ibid.*

34. *Ibid.*, p. 165.

35. For more on *k* see Rhodes, *Making of the Atomic Bomb*, p. 397.

36. Adsorption is a process whereby gases, liquids, or dissolved substances are gathered on a surface in a condensed layer.

37. Jones, *Manhattan*, p. 218.

38. Hewlett and Anderson, *New World*, p. 220.

39. Nichols, *Road to Trinity*, p. 108.

40. Rhodes, *Making of the Atomic Bomb*, p. 570.

41. *In the Matter of J. Robert Oppenheimer: Transcript of Hearing Before Personnel Security Board, Washington, D.C., April 12, 1954, Through May 6, 1954* (Washington, D.C.: Government Printing Office, 1954), pp. 12-13.

42. For details on Fuchs see Robert C. Williams, *Klaus Fuchs, Atom Spy* (Cambridge: Harvard University Press, 1987).

43. Rhodes, *Making of the Atomic Bomb*, pp. 540-41.

44. *Ibid.*, pp. 305-08.

45. *Ibid.*, pp. 644-45.

46. Jones, *Manhattan*, p. 530.

47. The Franck report, the response of the Scientific Panel, and the Interim Committee recommendations are discussed in *Ibid.*, pp. 365-72.

48. Rhodes, *Making of the Atomic Bomb*, pp. 571-72.

49. *Ibid.*, p. 676.

50. See *Ibid.*, pp. 668-78, for more on the Trinity test and the responses of those present.

51. Hewlett and Anderson, *New World*, p. 383.

52. *Ibid.*, p. 386.

53. Leslie R. Groves, *Now It Can Be Told*, (New York: Harper & Row, 1962), p. 434.

54. Herbert Feis, *The Atomic Bomb and the End of World War II* (Princeton: Princeton University Press, 1966), p. 85.

55. The directive is reprinted in Rhodes, *Making of the Atomic Bomb*, p. 691.

56. Hewlett and Anderson, *New World*, p. 395.

57. Paul W. Tibbets, "How to Drop an Atom Bomb," *Saturday Evening Post* 218 (June 8, 1946), p. 136.

58. The official yield of Little Boy, as listed in the Department of Energy's "Announced United States Nuclear Tests, July 1945 Through December 1986," page 2, January 1987, was 15,000 tons of TNT (15 kilotons).

59. The official yield of Fat Man listed in *Ibid*. Descriptions of the bombings of Hiroshima and Nagasaki derived from Rhodes, *Making of the Atomic Bomb*, pp. 710 and 739-40.

60. Summaries of Hiroshima and Nagasaki casualty rates and damage estimates appear in Rhodes, *Making of the Atomic Bomb*, pp. 733-34, 740, 742; Groves, *Now It Can Be Told*, pp. 319, 329-30, 346; and Jones, *Manhattan*, pp. 545-48.

61. The Smyth report is cited in footnote 25.

62. For details of the Baker fallout see Barton C. Hacker, *The Dragon's Tail: Radiation Safety in the Manhattan Project, 1942-1946* (Berkeley: University of California Press, 1987), pp. 135-53.

63. Hewlett and Anderson, *New World*, pp. 326-27.

64. *Ibid.*, p. 579.

65. Bundy, *Danger and Survival*, pp. 158-66.

66. The McMahon bill is reprinted in Hewlett and Anderson, *New World*, pp. 714-22.

67. The Manhattan Project ended with the transfer of power from the Manhattan Engineer District to the Atomic Energy Commission, though the Manhattan Engineer District itself was not abolished until August 15, 1947. The Top Policy Group and the Military Policy Committee had already ceased to exist by the time the Atomic Energy Commission took over on January 1, 1947.

Select Bibliography

Bundy, McGeorge, *Danger and Survival: Choices About the Bomb in the First Fifty Years* (New York: Random House, 1988).

Cantelon, Philip L. and Robert C. Williams, editors, *The American Atom: A Documentary History of Nuclear Policies from the Discovery of Fission to the Present, 1939-1984* (Philadelphia: University of Pennsylvania Press, 1984).

Groves, Leslie R., *Now It Can Be Told* (New York: Harper & Row, 1962).

Herken, Gregg, *Brotherhood of the Bomb: The Tangled Lives and Loyalties of Robert Oppenheimer, Ernest Lawrence and Edward Teller* (New York: Henry Holt, 2002).

Hersey, John, *Hiroshima* (New York: Knopf, 1946).

Hewlett, Richard G. and Oscar Anderson, Jr., *The New World, 1939-1946*, Volume I of A *History of the United States Atomic Energy Commission* (University Park: Pennsylvania State University Press, 1962).

Jones, Vincent C., *Manhattan: The Army and the Atomic Bomb* (Washington, D.C.: U. S. Government Printing Office, 1985).

Norris, Robert S., *Racing for the Bomb: General Leslie R. Groves, the Manhattan Project's Indispensable Man* (Hanover, NH: Steerforth Press, 2003).

Rhodes, Richard, *The Making of the Atomic Bomb* (New York: Simon & Schuster, 1986).

Smyth, Henry D., *Atomic Energy for Military Purposes: The Official Report on the Development of the Atomic Bomb Under the Auspices of the United States Government, 1940-1945* (Washington, D.C.: U. S. Government Printing Office, 1945).

United States Atomic Energy Commission, *In the Matter of J. Robert Oppenheimer: Transcript of Hearing Before Personnel Security Board, Washington, D.C., April 12, 1954, Through May 6, 1954* (Washington, D.C.: U. S. Government Printing Office, 1954).

Manhattan Project Chronology

Date	Events
1919	Ernest Rutherford discovers the proton by artificially transmuting an element (nitrogen into oxygen).
1930	Ernest O. Lawrence builds the first cyclotron in Berkeley.
1931	Robert J. Van de Graaff develops the electrostatic generator.
1932	James Chadwick discovers the neutron.
1932	J. D. Cockroft and E. T. S. Walton first split the atom.
1932	Lawrence, M. Stanley Livingston, and Milton White operate the first cyclotron.
1934	Enrico Fermi produces fission.
December 1938	Otto Hahn and Fritz Strassmann discover the process of fission in uranium.
December 1938	Lise Meitner and Otto Frisch confirm the Hahn-Strassmann discovery and communicate their findings to Niels Bohr.
January 26, 1939	Bohr reports on the Hahn-Strassman results at a meeting on theoretical physics in Washington, D. C.
August 2, 1939	Albert Einstein writes President Franklin D. Roosevelt, alerting the President to the importance of research on chain reactions and the possibility that research might lead to developing powerful bombs.
September 1, 1939	Germany invades Poland.
October 11-12, 1939	Alexander Sachs discusses Einstein's letter with President Roosevelt. Roosevelt decides to act and appoints Lyman J. Briggs head of the Advisory Committee on Uranium.
October 19, 1939	Roosevelt informs Einstein that he has set up a committee to study uranium.
October 21, 1939	The Uranium Committee meets for the first time.
November 1, 1939	The Uranium Committee recommends that the government purchase graphite and uranium oxide for fission research.

March 1940	John R. Dunning and his colleagues demonstrate that fission is more readily produced in the rare uranium-235 isotope, not the more plentiful uranium-238.		body to the Office of Scientific Research and Development.
		July 2, 1941	The British MAUD report concludes that an atomic bomb is feasible.
Spring-Summer 1940	Isotope separation methods are investigated.	**July 11, 1941**	A second National Academy of Sciences report confirms the findings of the first.
June 1940	Vannevar Bush is named head of the National Defense Research Committee. The Uranium Committee becomes a scientific subcommittee of Bush's organization.	**July 14, 1941**	Bush and Conant receive the MAUD report.
		October 9, 1941	Bush briefs Roosevelt and Vice President Henry A. Wallace on the state of atomic bomb research. Roosevelt instructs Bush to find out if a bomb can be built and at what cost. Bush receives permission to explore construction needs with the Army.
February 24, 1941	Glenn T. Seaborg's research group discovers plutonium.		
March 28, 1941	Seaborg's group demonstrates that plutonium is fissionable.		
May 1941	Seaborg proves plutonium is more fissionable than uranium-235.	**November 9, 1941**	A third National Academy of Sciences report agrees with the MAUD report that an atomic bomb is feasible.
May 17, 1941	A National Academy of Sciences report emphasizes the necessity of further research.	**November 27, 1941**	Bush forwards the third National Academy of Sciences report to the President.
June 22, 1941	Germany invades the Soviet Union.	**December 7, 1941**	The Japanese attack Pearl Harbor.
June 28, 1941	Bush is named head of the Office of Scientific Research and Development. James B. Conant replaces Bush at the National Defense Research Committee, which becomes an advisory	**December 10, 1941**	Germany and Italy declare war on the United States.
		December 16, 1941	The Top Policy Committee becomes primarily responsible for making

broad policy decisions relating to uranium research.

December 18, 1941 The S-l Executive Committee (which replaced the Uranium Committee in the Office of Scientific and Research Development) gives Lawrence $400,000 to continue electromagnetic research.

January 19, 1942 Roosevelt responds to Bush's November 27 report and approves production of the atomic bomb.

March 9, 1942 Bush gives Roosevelt an optimistic report on the possibility of producing a bomb.

May 23, 1942 The S-l Executive Committee recommends that the project move to the pilot plant stage and build one or two piles (reactors) to produce plutonium and electromagnetic, centrifuge, and gaseous diffusion plants to produce uranium-235.

June 1942 Production pile designs are developed at the Metallurgical Laboratory in Chicago.

June 17, 1942 President Roosevelt approves the S-l Executive Committee recommendation to proceed to the pilot plant

stage and instructs that plant construction be the responsibility of the Army. The Office of Scientific Research and Development continues to direct nuclear research, while the Army delegates the task of plant construction to the Corps of Engineers.

July 1942 Kenneth Cole establishes the health division at the Metallurgical Laboratory.

August 7, 1942 The American island-hopping campaign in the Pacific begins with the landing at Guadalcanal.

August 13, 1942 The Manhattan Engineer District is established in New York City, Colonel James C. Marshall commanding.

August 1942 Seaborg produces a microscopic sample of pure plutonium.

September 13, 1942 The S-l Executive Committee visits Lawrence's Berkeley laboratory and recommends building an electromagnetic pilot plant and a section of a full-scale plant in Tennessee.

September 17, 1942 Colonel Leslie R. Groves is appointed head of the Manhattan Engineer District. He is promoted to Brigadier General six days later.

September 19, 1942 Groves selects the Oak Ridge, Tennessee, site for the pilot plant.

September 23, 1942 Secretary of War Henry Stimson creates a Military Policy Committee to help make decisions for the Manhattan Project.

October 3, 1942 E. I. du Pont de Nemours and Company agrees to build the chemical separation plant at Oak Ridge.

October 5, 1942 Compton recommends an intermediate pile at Argonne.

Fall 1942 J. Robert Oppenheimer and the luminaries report from Berkeley that more fissionable material may be needed than previously thought.

October 19, 1942 Groves decides to establish a separate scientific laboratory to design an atomic bomb.

October 26, 1942 Conant recommends dropping the centrifuge method.

November 12, 1942 On the recommendation of Groves and Conant, the Military Policy Committee decides to skip the pilot plant stage on the plutonium, electromagnetic, and gaseous diffusion projects and go directly from the research stage to industrial-scale production. The Committee also decides not to build a centrifuge plant.

November 14, 1942 The S-1 Executive Committee endorses the recommendations of the Military Policy Committee.

November 1942 The Allies invade North Africa.

November 25, 1942 Groves selects Los Alamos, New Mexico, as the bomb laboratory (codenamed Project Y). Oppenheimer is chosen laboratory director.

December 2, 1942 Scientists led by Enrico Fermi achieve the first self-sustained nuclear chain reaction in Chicago.

December 10, 1942 The Lewis committee compromises on the electromagnetic method. The Military Policy Committee decides to build the plutonium production facilities at a site other than Oak Ridge.

December 28, 1942 Roosevelt approves detailed plans for building production facilities and producing atomic weapons.

January 13-14, 1943 Plans for the Y-12 electromagnetic plant are discussed. Groves insists that Y-12's first racetrack be finished by July 1.

January 14-24, 1943   At the Casablanca Conference, Roosevelt and British Prime Minister Churchill agree upon unconditional surrender for the Axis powers.

January 16, 1943   Groves selects Hanford, Washington, as the site for the plutonium production facilities. Eventually three reactors, called B, D, and F, are built at Hanford.

January 1943   Bush encourages Philip Abelson's research on the thermal diffusion process.

February 18, 1943   Construction of Y-12 begins at Oak Ridge.

February 1943   Groundbreaking for the X-10 plutonium pilot plant takes place at Oak Ridge.

March 1943   Researchers begin arriving at Los Alamos.

April 1943   Bomb design work begins at Los Alamos.

June 1943   Site preparation for the K-25 gaseous diffusion plant commences at Oak Ridge.

Summer 1943   The Manhattan Engineer District moves its headquarters to Oak Ridge.

July 1943   Oppenheimer reports that three times as much fissionable material may be necessary than thought nine months earlier.

August 27, 1943   Groundbreaking for the 100-B plutonium production pile at Hanford takes place.

September 8, 1943   Italy surrenders to Allied forces.

September 9, 1943   Groves decides to double the size of Y-12.

September 27, 1943   Construction begins on K-25 at Oak Ridge.

November 4, 1943   The X-10 pile goes critical and produces plutonium by the end of the month.

Late 1943   John von Neumann visits Los Alamos to aid implosion research.

December 15, 1943   The first Alpha racetrack is shut down due to maintenance problems.

January 1944   The second Alpha racetrack is started and demonstrates maintenance problems similar to those that disabled the first.

January 1944   Construction begins on Abelson's thermal diffusion plant at the Philadelphia Naval Yard.

February 1944   Y-12 sends 200 grams of uranium-235 to Los Alamos.

March 1944   The Beta building at Y-12 is completed.

March 1944	Bomb models are tested at Los Alamos.	**September 1944**	Colonel Paul Tibbets's 393rd Bombardment Squadron begins test drops with dummy bombs called pumpkins.
April 1944	Oppenheimer informs Groves about Abelson's thermal diffusion research in Philadelphia.	**September 13, 1944**	The first slug is placed in pile 100-B at Hanford.
June 6, 1944	Allied forces launch the Normandy invasion.	**September 1944**	Roosevelt and Churchill meet in Hyde Park and sign an "aide-memoire" pledging to continue bilateral research on atomic technology.
June 21, 1944	Groves orders the construction of the S-50 thermal diffusion plant at Oak Ridge.		
July 4, 1944	The decision is made to work on a calutron with a 30-beam source for use in Y-12.	**Summer 1944-Spring 1945**	The Manhattan Project's chances for success advance from doubtful to probable as Oak Ridge and Hanford produce increasing amounts of fissionable material, and Los Alamos makes progress in chemistry, metallurgy, and weapon design.
July 17, 1944	The plutonium gun bomb (codenamed Thin Man) is abandoned.		
July 1944	A major reorganization to maximize implosion research occurs at Los Alamos.	**September 27, 1944**	The 100-B reactor goes critical and begins operation.
July 1944	Scientists at the Metallurgical Laboratory issue the "Prospectus on Nucleonics," concerning the international control of atomic energy.	**September 30, 1944**	Bush and Conant advocate international agreements on atomic research to prevent an arms race.
		December 1944	The chemical separation plants (Queen Marys) are finished at Hanford.
August 7, 1944	Bush briefs General George C. Marshall, informing him that small implosion bombs might be ready by mid-1945 and that a uranium bomb will almost certainly be ready by August 1, 1945.	**February 2, 1945**	Los Alamos receives its first plutonium.

February 4-11, 1945   Roosevelt, Churchill, and
                       Soviet Premier Joseph
                       Stalin meet at Yalta.

March 1945   S-50 begins operation at
                       Oak Ridge.

March 1945   Tokyo is firebombed,
                       resulting in 100,000
                       casualties.

March 12, 1945   K-25 begins production at
                       Oak Ridge.

April 12, 1945   President Roosevelt dies.

April 25, 1945   Stimson and Groves brief
                       President Truman on the
                       Manhattan Project.

May 1945   Stalin tells Harry Hopkins
                       that he is willing to meet
                       with Truman and proposes
                       Berlin as the location.

May 7, 1945   The German armed forces
                       in Europe surrender to the
                       Allies.

May 23, 1945   Tokyo is firebombed again,
                       this time resulting in
                       83,000 deaths.

May 31-June 1, 1945   The Interim Committee
                       meets to make
                       recommendations on
                       wartime use of atomic
                       weapons, international
                       regulation of atomic
                       information, and legislation
                       regarding domestic control
                       of the atomic enterprise
                       (the Committee's draft

legislation becomes the
basis for the May-Johnson
bill).

June 6, 1945   Stimson informs President
                       Truman that the Interim
                       Committee recommends
                       keeping the atomic bomb
                       a secret and using it as
                       soon as possible without
                       warning.

June 1945   Scientists at the
                       Metallurgical Laboratory
                       issue the Franck Report,
                       advocating international
                       control of atomic
                       research and proposing
                       a demonstration of the
                       atomic bomb prior to its
                       combat use.

June 14, 1945   Groves submits the
                       target selection group's
                       recommendation to
                       Marshall.

June 21, 1945   The Interim Committee,
                       supporting its Scientific
                       Panel, rejects the Franck
                       Report recommendation
                       that the bomb be
                       demonstrated prior to
                       combat.

July 2-3, 1945   Stimson briefs Truman on
                       the Interim Committee's
                       deliberations and outlines
                       the peace terms for Japan.

July 16, 1945   Los Alamos scientists
                       successfully test a

	plutonium implosion bomb in the Trinity shot at Alamogordo, New Mexico.	**August 9, 1945**	The implosion model plutonium bomb, called Fat Man, is dropped on Nagasaki.
July 17- August 2, 1945	Truman, Churchill, and Stalin meet in Potsdam.	**August 12, 1945**	The Smyth Report, containing unclassified technical information on the bomb project, is released.
July 21, 1945	Groves sends Stimson a report on the Trinity test.		
July 24, 1945	Stimson again briefs Truman on the Manhattan Project and peace terms for Japan. In an evening session, Truman informs Stalin that the United States has tested a powerful new weapon.	**August 14, 1945**	Japan surrenders.
		September 2, 1945	The Japanese sign articles of surrender aboard the U.S.S. *Missouri*.
		September 9, 1945	S-50 shuts down.
July 25, 1945	The 509th Composite Group is ordered to attack Japan with an atomic bomb "after about" August 3.	**September 1945**	Y-12 shutdown begins.
		October 3, 1945	Truman advocates passage of the May-Johnson bill.
July 26, 1945	Truman, Chinese President Chiang Kai-Shek, and new British Prime Minister Clement Atlee issue the Potsdam Proclamation, calling for Japan to surrender unconditionally.	**December 20, 1945**	Senator Brien McMahon introduces a substitute to the May-Johnson bill, which had been losing support, including Truman's.
July 29, 1945	The Japanese reject the Potsdam Proclamation.	**January 1946**	Hearings on the McMahon bill begin.
August 6, 1945	The gun model uranium bomb, called Little Boy, is dropped on Hiroshima. Truman announces the raid to the American public.	**June 14, 1946**	Bernard Baruch presents the American plan for international control of atomic research.
August 8, 1945	Russia declares war on Japan and invades Manchuria.	**July 1, 1946**	Operation Crossroads begins with Shot Able, a plutonium bomb dropped from a B-29, at Bikini Atoll.

July 25, 1946	Operation Crossroads continues with Shot Baker, a plutonium bomb detonated underwater, at Bikini Atoll.
August 1, 1946	President Truman signs the Atomic Energy Act of 1946, a slightly amended version of the McMahon bill.
December 1946-January 1947	The Soviet Union opposes the Baruch Plan, rendering it useless.
January 1, 1947	In accordance with the Atomic Energy Act of 1946, all atomic energy activities are transferred from the Manhattan Engineer District to the newly created United States Atomic Energy Commission. The Top Policy Group and the Military Policy Committee had already disbanded.
August 15, 1947	The Manhattan Engineer District is abolished.
December 31, 1947	The National Defense Research Committee and the Office of Scientific Research and Development are abolished. Their functions are transferred to the Department of Defense.